Synthesis Lectures on Computer Science

The series publishes short books on general computer science topics that will appeal to advanced students, researchers, and practitioners in a variety of areas within computer science.

Jeffrey W. Herrmann

Metareasoning for Robots

Adapting in Dynamic and Uncertain Environments

 Springer

Jeffrey W. Herrmann
University of Maryland
College Park, MD, USA

ISSN 1932-1228   ISSN 1932-1686 (electronic)
Synthesis Lectures on Computer Science
ISBN 978-3-031-32239-6   ISBN 978-3-031-32237-2 (eBook)
https://doi.org/10.1007/978-3-031-32237-2

This Springer imprint is published by the registered company Springer Nature Switzerland AG
The registered company address is: Gewerbestrasse 11, 6330 Cham, Switzerland

To L.G.H. and C.R.H.

Preface

This book covers metareasoning, a branch of artificial intelligence (AI) that is increasingly being used to improve the performance of autonomous systems, including mobile ground robots and unmanned aerial systems. It improves performance by working behind the scenes to avoid (or recover from) reasoning failures, select the best algorithm, or otherwise optimize the use of computational resources. In some ways, metareasoning resembles that old BASF commercial, "We don't make a lot of the products you buy. We make a lot of the products you buy better." Metareasoning includes a variety of beneficial techniques that have been implemented in many ways, so an engineer who is interested in using metareasoning might be overwhelmed. Although many papers describing metareasoning approaches have been published, there is no single place to find them. Previous books on metareasoning have emphasized state-of-the-art research results but have not explained how to create a metareasoning approach.

We have reached the point where we can begin to offer concrete advice on how to develop metareasoning, however. Certainly, there is not yet a single way to implement metareasoning to improve a robot's performance, and I expect that such an approach is infeasible given the diversity of robots that engineers and organizations are developing for an incredible range of applications.

Still, enough has been done that we can learn from those efforts. This book presents a systems engineering process for developing a metareasoning approach. It is based on the work that I, my colleagues and students, and others have done to develop and implement metareasoning approaches. The book uses their work as examples to illustrate the metareasoning engineering process.

The overall structure of the book follows this process for developing, implementing, and testing a metareasoning approach. By going through these steps, an engineer can systematically consider the metareasoning alternatives and make decisions about the metareasoning approach.

Chapter 1 starts our study by considering why we might use metareasoning on a robot and what metareasoning is. This chapter introduces some key concepts related to robots and autonomous systems, discusses the benefits of metareasoning, and presents a list of key sources that one should read for more information about metareasoning. Finally,

Chap. 1 introduces the metareasoning engineering process that informs the structure and contents of this book.

Chapter 2 reviews the metareasoning design options. Making these decisions begins to specify how the robot will perform metareasoning. A key decision is to select the reasoning processes that the meta-level should control. This chapter discusses the options that are available for metareasoning approaches and policies, using examples from our own work and the literature to illustrate the techniques, advantages, and drawbacks of different options.

Chapter 3 describes the options for locating the software that will implement the metareasoning process on a robot. Because the implementation details depend upon the robot's autonomy software architecture, this chapter reviews some common architectures. This chapter also presents a specific implementation on a mobile ground robot.

Chapter 4 describes a systematic, data-driven approach for determining the metareasoning policy and presents some case studies to illustrate the synthesis process. These case studies provide insights that engineers can use in their own work on metareasoning.

Chapter 5 discusses testing, which is important for determining whether and how much metareasoning improves the performance of the robot. It is also useful for evaluating different metareasoning policies to see which one performs best. This chapter describes different performance metrics, test planning, ways to visualize metareasoning, data analysis techniques, and assurance cases.

Thus, this concise volume gives an engineer a roadmap, to speak, on how to proceed with metareasoning. Although it does not provide a metareasoning "solution" that will be useful for everyone, one can follow the steps of the process (with some iteration as needed) to develop, implement, and test a metareasoning approach for a specific application.

This book is designed for engineers, roboticists, computer scientists, students, and others who wish to improve the performance of their robots with metareasoning. The chapters refer to a variety of technologies and algorithms that are used to develop robots; where possible, they provide links to references where readers can learn more about those topics. They also cite papers that describe previous work on metareasoning approaches. Each chapter has a list of references cited.

College Park, MD, USA   Jeffrey W. Herrmann

Acknowledgements

My study of metareasoning has been supported by various organizations, including the Air Force Research Laboratory, the Naval Air Warfare Center-Aircraft Division, and the Army Research Laboratory. I have benefitted from my appointment to the Institute for Systems Research at the University of Maryland and the resources of the Maryland Robotics Center. I have been greatly assisted and encouraged by my colleagues, including Shapour Azarm, Michael Otte, Huan (Mumu) Xu, and Kevin Carey, and our students and have enjoyed collaborating with them.

Robert St. Amant, Michael Dawson, and Sidney Molnar read and provided valuable feedback on a draft version of this manuscript. Some of the material in this book previously appeared in various papers that I and my collaborators have written, but all of the chapters are original.

I also gratefully acknowledge the editorial and production staff of Springer Nature for their efforts to develop and publish this book.

Finally, I want to express my appreciation for my wife's constant support and encouragement throughout my career.

Contents

About the Author

Jeffrey W. Herrmann is a professor at the University of Maryland, where he holds a joint appointment with the Department of Mechanical Engineering and the Institute for Systems Research. He is a member of the Society of Catholic Scientists, IISE, ASME, and the Design Society. In 2012, he and Gino Lim were the conference chairs for the Industrial and Systems Engineering Research Conference.

Dr. Herrmann earned his B.S. in applied mathematics from the Georgia Institute of Technology. As a National Science Foundation Graduate Research Fellow from 1990 to 1993, he received his Ph.D. in industrial and systems engineering from the University of Florida. His dissertation investigated production scheduling problems motivated by semiconductor manufacturing. He held a post-doctoral research position at the Institute for Systems Research from 1993 to 1995.

Dr. Herrmann's research, service, and teaching activities have established him as a leader in the following areas: (1) developing novel mathematical models to improve public health preparedness, (2) describing and modeling engineering design decision-making processes, and (3) using risk-based techniques to improve path planning for autonomous systems. Through February 2023, he has published over 140 journal papers, refereed conference papers and fifteen book chapters, co-authored an engineering design textbook, edited two handbooks, and authored a textbook on engineering decision-making and risk management. Dr. Herrmann and his colleagues have been awarded over $34 million in research contracts and grants, including research awards from NSF, Amazon, the Naval Air Warfare Center, the Air Force Research Laboratory, and the Army Research Laboratory. He has advised over 50 Ph.D. dissertations and M.S. theses. In 2003, Dr. Herrmann received the Society of Manufacturing Engineers Jiri Tlusty Outstanding Young Manufacturing Engineer Award; in 2013, he was named a Diplomate of the Society for Health Systems. In 2016, his textbook won the IIE/Joint Publishers Book of the Year award.

Introduction to Metareasoning

Despite the promises (and threats) of science fiction stories, robots (and other autonomous systems) are not very intelligent, and they sometimes fail. In the real world, robots run into obstacles and other robots, get stuck in corners, wander away aimlessly, stop suddenly for no reason, pick up the wrong object, drop things, fall over, crash into the ground, and do many other undesirable or unsafe things.

A robot can do only certain activities and behaviors, which are limited by its programming and its mechanical subsystems. Its "brain" executes computer programs that generate and store (or send) outputs in response to their inputs. When a robot "thinks" or "reasons," it is running an algorithm that computes values, manipulates numbers and symbols (such as words and logical values), executes other algorithms, or searches for a solution to a well-posed problem. Some robots have programs that "learn" by adjusting the values in lookup tables or the coefficients of formulas so that those calculations will be more accurate in the future. All of the robot's mechanisms, algorithms, and programs are designed by humans who want the robot to perform specific activities and behaviors safely and reliably.

Anyone who is designing a robot should keep in mind the limited resources (e.g., computational resources, memory, and power) that are available for performing the reasoning processes that will enable the robot to sense its environment and make rational decisions about what to do next. To make our robot perform better, we want more resources, but adding more resources makes the robot larger, more massive, more powerful, and more expensive, which is a big problem. We need another approach. Is there a way that we can optimize the use of the robot's resources (instead of adding more)? To do more with less?

This book is for engineers and programmers who wish to design a better robot. It describes metareasoning, a type of artificial intelligence (AI) that one can use to improve the robot's ability to reason in the face of its limited resources and in a dynamic, uncertain environment. Metareasoning can select the best planning algorithm, recover from a

© The Author(s), under exclusive license to Springer Nature Switzerland AG 2023
J. W. Herrmann, *Metareasoning for Robots*, Synthesis Lectures on Computer Science,
https://doi.org/10.1007/978-3-031-32237-2_1

reasoning process failure, and adjust the parameters of a reasoning algorithm. Although metareasoning has been discussed for many years, only now are engineers and programmers beginning to use it to design more intelligent robots that are safer and more reliable.

This chapter introduces some key concepts related to robots and autonomous systems, discusses why metareasoning is needed and the benefits of metareasoning, and presents a list of key sources that one should read for more information about metareasoning. Finally, it discusses the systems engineering approach that informs the structure and contents of this book.

1.1   Key Concepts

1.1.1   Robot

This book focuses on robots and similar systems that sense, reason, and act. A "robot" is a "programmed actuated mechanism with a degree of autonomy to perform locomotion, manipulation or positioning" (ISO 2021). An "intelligent robot" is "a physically situated intelligent agent," according to Murphy (2019), which means that it has a physical existence, can perceive its environment, and take actions to accomplish a mission. A robot is not a purely computational agent that can only perform computations and output the results of those computations (Zilberstein 1993).

In 2023 commercially available robots include the Roomba vacuum cleaner, the Amazon Astro home monitoring device, and the Automower robotic lawn mower. Industrial robots include those used on automotive assembly lines for welding and other operations and material handling robots such as the drive units that operate in Amazon's distribution centers.

Any unmanned vehicle is a type of robot. Unmanned aerial vehicles (commonly known as "drones"), such as those made by Skydio, are a special class of robots. One outstanding example is the X-37B, a spacecraft that can spend hundreds of days in low Earth orbit while performing scientific experiments and classified missions for the U.S. Air Force. At the end of a mission, the X-37B can autonomously de-orbit and land (Boeing 2022).

The National Aeronautics and Space Administration (NASA) has sent robots ("rovers") to Mars, including *Opportunity*, *Spirit*, *Sojourner* (shown in Fig. 1.1), *Perseverance*, and *Curiosity*. Due to the distance from Earth and the Mars and the resulting lag in communications, it is not feasible for someone on Earth to operate the rovers remotely; thus, they must be able to reason and make decisions autonomously.

Fig. 1.1 The Sojourner rover on Mars, as seen by the Pathfinder lander (*Photo credit* National Aeronautics and Space Administration; *Image source* https://en.wikipedia.org/wiki/File:Sojourner_on_Mars_PIA01122.jpg)

1.1.2 Autonomy

"Autonomy" is the "ability to perform intended tasks based on current state and sensing, without human intervention" (ISO 2021). Murphy (2019) identified four aspects of autonomy: generating plans, performing non-deterministic actions, using open world models, and representing knowledge as symbols. Usually, a robot operates by executing a plan, and an autonomous robot can generate its own plans, instead of relying on a human to give it a plan. Moreover, because an autonomous robot uses noisy sensors, multiple computational processes, and unreliable hardware and operates in a dynamic environment, it will likely exhibit non-deterministic (unpredictable) behavior. Although users might view this uncertain behavior as a drawback, it is essential for giving the robot the ability to adapt to its environment and other changes. An autonomous robot uses an open world model that describes the environment, but the robot does not assume that the world model describes every possible object, state, or condition. This gives the robot the ability to operate successfully in a wider range of situations. Finally, an autonomous system can manipulate symbols, in addition to signals, to enable more powerful reasoning processes.

There are different levels of autonomy, and some organizations have identified these explicitly. For example, the U.S. Air Force (Office of the Chief Scientist 2015) described the following levels of autonomy (each of which gives the system more capabilities):

1. Fully manual: the human operator completes all parts of the task;
2. Implementation aiding: the human operator makes all of the decisions, and the system merely completes the tasks for the operator;
3. Situation awareness support: the system can integrate data and provide information to the operator;

4. Decision aiding: the system can score or rank the options;
5. Supervisory control: the human operator determines the system's goals and intervenes if necessary, but the system decides which actions to perform; and
6. Full autonomy: the system performs the task without human supervision or assistance.

A Roomba can move around a room without human intervention, so it can perform that task with full autonomy. My car's navigation system can tell me the fastest way to get home, but it can't drive the car, so it is a decision aiding system.

1.1.3   Reasoning

A robot's reasoning processes are performed by algorithms that run on the robot's computer. These algorithms support sensing and perception, localization and mapping, navigation, motion planning, and other reasoning activities. Reasoning algorithms use information from sensors and other sources and generate commands that tell the robot's actuators what to do. For example, a robot is reasoning when its computer processes the signals from its LIDAR, detects an obstacle in its way, decides to move to the right, and send commands to the motors that drive its wheels. A reasoning process "decides" something when the corresponding algorithm calculates or sets the value of the variables that determine what the robot will do.

A reasoning process executes an algorithm to perform a computational function in response to its inputs, stores or send its output, and then repeats this in response to new inputs. The algorithm that a reasoning process executes might be a very simple function or a sequence of subroutines, each of which implements one or more algorithms. A reactive system (such as a Roomba) can operate without planning (Murphy 2019), but that system still has an algorithm (possibly a very simple function) that transforms the sensor inputs into a command.

1.1.4   Rationality

Ideally, when it needs to reason, a robot would use appropriate algorithms that always generate the best possible solutions (the ones that lead to optimal performance). Although this is desirable, it is not feasible, because many of the problems that need to be solved are computationally complex, and generating optimal solutions would require excessive time and computational resources for real-time environments. Unfortunately, every robot has finite size, weight, and power, which limits the computational resources that can be used, and a robot needs to reason and act within a reasonable amount of time in order to accomplish its mission. Executing a reasoning process requires time, and it might lead to the robot selecting an action that was feasible and desirable when the reasoning process began but is not by the time it was selected (when the reasoning process ended).

Thus, because a perfectly rational robot or autonomous agent is not achievable, we must design robots that have bounded rationality (Svegliato and Zilberstein 2018), and this sometimes requires making difficult choices about which algorithms to deploy on a robot.

Bounded rationality describes the capabilities of a decision-maker in the presence of limited time, finite computational resources, and other resource constraints (Simon 1981). A decision-making system that tried to find an optimal choice in every situation would need excessive amounts of computational resources and would still require a long time to do so.

Good (1983) identified two types of rationality: "type I" rationality emphasizes finding the best alternative (the one with the greatest utility) without considering the time and cost of the computations needed to do so, and "type II" rationality considers the cost (negative utility) of searching as well as the benefit (positive utility) of the alternative is selected. This is also known as *optimization under constraints*. Although the type II approach appears to be more reasonable, it still requires solving a difficult optimization problem. For that reason, some researchers consider that type II rationality is not actually bounded rationality (Gigerenzer and Selten 2001).

This leads us to consider heuristics, which form an important class of decision-making procedures. Heuristics are rules or algorithms that do not seek to find an optimal alternative (or solution). A heuristic will sometimes find an optimal solution, but that is not guaranteed. A simple decision rule can select an alternative based on one relevant factor. A solution algorithm can construct a solution using a small number of policies (instead of exhaustively searching all possible solutions). When the available computational resources are limited, using appropriate heuristics can perform well, leading to a different type of bounded rationality (Gigerenzer and Selten 2001). The debate over bounded rationality has lasted years and yielded some useful insights, but it leads to more theoretical aspects, so we won't pursue it here.

In general, from a practical point of view, we face a persistent, sometimes frustrating, tradeoff: some algorithms generate better solutions but require more computational effort, while others need less computational effort but might generate inferior solutions. Moreover, the algorithms that are deployed might have numerous parameters that affect their performance. Finally, no algorithm is ideal for every situation that the robot might encounter. All of these factors make the design decisions very complicated.

1.1.5 Metareasoning

Metareasoning is the process of reasoning about reasoning (Cox and Raja 2011; Genesereth and Nilsson 1987; Russell and Wefald 1991a, b). The following statement describes a fundamental principle about metareasoning: *Metareasoning monitors and controls a robot's reasoning processes.* Metareasoning enables bounded rationality by giving the robot the ability to reason about its own reasoning process and to consider the cost and time of computation as it considers which action to perform next (Svegliato and Zilberstein 2018). In principle, a robot can use metareasoning to allocate its scarce computational resources optimally. As we'll see, however, this goal is beyond the scope of existing metareasoning approaches because the problem is NP-complete in the strong sense (Zilberstein and Russell 1995). (An NP-complete problem is inherently intractable, so no algorithm can solve the problem in a reasonable amount of time. The theory of NP-completeness provides ways to prove that a complex problem is inherently intractable; Garey and Johnson 1979.)

Metareasoning is a formalization of metacognition, which an intelligent agent does when it thinks about its own thinking (Cox 2005). Metareasoning treats computations as actions and evaluates a computation based on (a) the time (and other resources) that are required to execute it and (b) its impact on the robot's actions in the real world (Russell and Wefald 1991b). An agent uses metareasoning, also known as metalevel control, to improve the quality of its decisions. Examples include determining which algorithm to use to make a decision and determining when to stop computing and execute an action.

Anderson and Oates (2007), Cox (2005), Cox and Raja (2011), and Russell and Wefald (1991a, b) presented fundamental concepts in metareasoning. The canonical three-level model of metareasoning (Cox and Raja 2011) organizes a robot's actions as follows (cf. Fig. 1.2):

- the *ground level* contains the robot's interactions with the environment (including motion and sensing);
- the *object level* contains the robot's reasoning processes, which represent and use information from the robot's sensors, select appropriate actions to complete a task, and command the robot to perform those actions; and
- the *meta-level* contains the robot's metareasoning approach, which monitors and controls the reasoning processes in the object level.

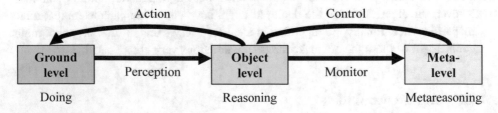

Fig. 1.2 The three-level model of metareasoning, including the ground level, the object level, and the meta-level (adapted from Cox and Raja 2011)

That is, the object level contains the algorithms and other procedures that form the robot's decision-making process through deliberation and means-ends reasoning; these can be organized into multiple layers (Parashar and Goel 2021). This decision-making process determines which, when, and how ground-level actions should be performed to achieve the robot's goals. For example, in a robot's software stack, the image processing, simultaneous localization and mapping (SLAM), and planning algorithms are at the object level.

As shown in Fig. 1.2, the robot's decision-making process (in the object level) thinks about and controls the robot's actions (in the ground level), and metareasoning (in the meta-level) thinks about and controls the robot's reasoning (in the object level). Different terms have been used for these levels. For instance, Brachman (2002) used the term "reflective" to describe the meta-level, and Svegliato et al. (2019) used the term "introspective."

Of course, Fig. 1.2 is an abstraction that oversimplifies the situation. A robot might have many different sensors and actuators, and a robot's decision-making process typically uses a variety of algorithms that interact in many ways. That is, there might be multiple reasoning processes in the object level. Thus, the meta-level might be a collection of metareasoning policies, each of which monitors and controls one part (e.g., one algorithm) of the object level's decision-making process (Parashar and Goel 2021).

Naturally, if metareasoning is reasoning about reasoning, then a robot can reason about metareasoning (meta-metareasoning), reason about meta-metareasoning (meta-meta-metareasoning) and so forth. This would lead to an infinite regress, which is certainly impossible to execute in a real-time system. In a bounded rational robot, infinite regress would prevent any computations from being completed. Thus, as the robot's designers, we must determine which computations the robot will complete without reasoning about them in real time. One solution is that the robot will perform metareasoning but not meta-metareasoning. The metareasoning computations are not monitored and controlled in real time; they are limited to the computations needed to execute the metareasoning policy that the engineer has selected.

Metareasoning is a branch of AI that has been discussed for many years. Genesereth and Nilsson (1987) described metareasoning as a process of inference on a database of statements about the reasoning procedure. They defined "reflection" as the process of (a) suspending the current reasoning process, (b) reasoning about that reasoning process, and (c) using the results to control subsequent reasoning. In particular, they described multi-level inference procedures that have causal connections between the reasoning at different levels. In one of their examples, the object level asks the meta-level which operation it should perform next, and the meta-level reasoning answers that query using its knowledge about the object level reasoning, which they call "metaknowledge."

This book discusses metareasoning for robots, which is a subset of the general topic. Metareasoning for robots must cope with the real-time nature of robots and the robot's limited computational resources. This book does not present a general-purpose framework for making optimal metareasoning decisions (cf. Russell and Wefald 1991a, b); instead, we'll focus on how one designs a metareasoning approach for a robot (see also Sect. 1.4).

Metareasoning is a large area and is related to other areas (such as metalearning and metacognition) that are beyond the scope of this book. This book does not present a systematic survey of all metareasoning research; instead, it describes the monitoring and control of reasoning within one robot and discusses metareasoning for a team of robots. Section 1.3 provides some key references that provide more information about metareasoning. Metareasoning is also considered in computational logic (Costantini 2002) and in human decision making (Ackerman and Thompson 2017; Cox 2005), but these are also outside the scope of this book. Finally, this book does not discuss dynamic resource allocation methods and resource-aware computing approaches that control computing resources such as a Multi-Processor System on Chip (MPSoC) (Paul et al. 2014).

1.2   Benefits of Metareasoning

As the other chapters of this book discuss, designing, implementing, and testing a metareasoning approach for a robot requires time and effort. *Is it worth it?* This section discusses why an engineer should consider metareasoning (Sect. 1.6 will cover some of the associated drawbacks).

Using metareasoning delays some design decisions about the reasoning system. We don't have to select and implement one algorithm (which likely performs well in some situations and poorly in others). Instead, we delegate decisions to the metareasoning approach, and thus "the nature and order of computation steps in the object-level decision procedure are left open to events occurring after the system is created" (Russell and Wefald 1991a). For instance, Siegwart et al. (2011) described a variety of obstacle avoidance algorithms that a robot might use and listed their characteristics. No single algorithm is ideal for every situation. Instead of committing to a single algorithm, we can implement a metareasoning approach that selects the best one in real time. Using metareasoning to control anytime algorithms, instead of setting a fixed time limit for computation, also defers a design decision (Zilberstein 1993).

Next, due to this flexibility, metareasoning can improve performance. This is most notable in algorithm selection, where the metareasoning policy seeks to select the algorithm that will yield the best solution for the current problem instance. For example, the SATzilla program, created to solve propositional satisfiability (SAT) problems, used algorithm selection to win gold medals in the 2007 and 2009 SAT competitions; moreover, "the solvers contributing most to SATzilla were often not the overall best-performing solvers, but instead solvers that exploit novel solution strategies to solve instances that would remain unsolved without them" (Xu et al. 2012). Likewise, a metareasoning policy that dynamically schedules reasoning processes instead of using a fixed cyclic schedule with worst-case frequencies led to better performance of an avionics system (Greenwald

1996). Using a metareasoning policy that dynamically changed the task allocation algorithm reduced the time for a multi-agent system to complete a search mission (Carrillo et al. 2021). The metareasoning policy switched the task allocation algorithm when the communication availability changed. Boddy and Dean (1994) discussed a shooting gallery robot with a meta-level that used metareasoning to determine when to schedule the robot's actions (instead of using a simple rule). Their experiments showed that using metareasoning increased the robot's score by approximately 20%. Molnar et al. (2023) studied an autonomous mobile ground robot with a meta-level that switched its path planning algorithm whenever the currently active algorithm was unable to find a feasible solution or took too long. The results of their simulation tests showed that using metareasoning reduced the frequency of mission failures due to path planning failures.

In addition, metareasoning gives a robot the ability to deal with unexpected situations, which can also make the robot safer. Metareasoning complements reinforcement learning, social learning, and using analogies as strategies for addressing novel situations (Goel et al. 2020). Metareasoning is a type of reflection based on self-awareness, and a self-aware, reflective system is more robust because it can modify a reasoning process that has led (or would lead) a robot into an undesirable situation (Anderson and Oates 2007; Brachman 2002). For instance, Svegliato et al. (2019) described a metareasoning approach (called *introspection*) that interrupts an autonomous vehicle's regular reasoning process whenever an exception handler indicates that an abnormal situation is occurring (e.g., a garbage truck blocking a two-lane road). It then invokes a different decision process that is appropriate for that abnormal situation.

In their discussion of the results of a workshop on robot planning research, Alterovitz et al. (2016) identified the need to utilize available resources effectively and the ability to adapt to changes in the environment as two desirable capabilities for robots. By monitoring and controlling object-level reasoning, metareasoning enables these capabilities. Although the performance and cost of computational resources (such as CPUs and GPUs) is improving over time, at some point we must select specific chips and boards for our robot, and then we need to determine the best way to use them.

Finally, metareasoning is a promising approach for achieving *bounded rationality*, which describes an agent's ability to make good decisions despite limited time and resources for deliberation (Simon 1981; Svegliato and Zilberstein 2018). For instance, the Amazon Astro robot is a small, inexpensive robot for household use. Its small size is desirable for navigating a room with many obstacles, and its low cost allows Amazon to sell them at a reasonable price. To achieve these characteristics, however, Amazon's engineers had to choose less powerful processors and less memory, which constrains the robot's computational power. Metareasoning is meant to improve (ideally, optimize) the performance of any robot that has limited computational resources (Horvitz 2001).

1.3   Metareasoning Reading List

Although this book provides the information needed to design a metareasoning approach for a robot, it is not a comprehensive discussion of metareasoning. Readers who wish to learn more about the background of this topic and details about the different approaches that have been developed might consider the following books and articles (and the works that they cite). This list is meant to get one started; it is not comprehensive.

Horvitz (1987). In this paper Horvitz analyzed the problem of anytime algorithms, considered the tradeoff between the time of computation and solution quality, and used classical decision theory to find an optimal solution. Horvitz wrote many other papers about metareasoning, and many are cited by Horvitz (2001).

Russell and Wefald (1991a). The first two chapters of this book discuss the limits that prevent robots from achieving perfect rationality and classify metareasoning architectures. The third chapter presents a model for the value of a computation. See also Russell and Wefald (1991b).

Boddy and Dean (1994). Boddy and Dean also used metareasoning to improve computation when the time available for computation is limited. They considered anytime algorithms and proposed deliberation scheduling, which determines ahead of time how much computational time should be allocated to the tasks that need to be completed in order to maximize the total value of the computations.

Zilberstein and Russell (1995). Zilberstein and Russell modeled the problem of controlling an algorithm that is a composition of multiple modules that are anytime algorithms. The metareasoning decision must determine how much computation time to allocate to each module. They applied their model to a robot navigation algorithm that includes a vision module and a planning algorithm.

Costantini (2002). This review paper emphasizes computational logic and logic programming, which demonstrates the broad applicability of metareasoning.

Cox (2005). This review paper covers the general area of metacognition, including human psychology, problem solving, artificial intelligence, metareasoning, and learning.

Anderson and Oates (2007). This review paper describes two broad areas of metareasoning research: (1) scheduling and controlling deliberation and (2) generating and using higher-order knowledge about reasoning processes. The work in the first area includes the theory of bounded optimality and controlling anytime algorithms (see Chap. 2 of this book). The authors also considered metalearning (such as learning to choose a learning algorithm).

Cox and Raja (2011). This edited book is an important collection of papers about metareasoning. The chapter by Cox and Raja is a fine overview of metareasoning concepts. The chapter by Goel and Jones has numerous references to early work on metareasoning and discusses the relationships between metareasoning, self-adaptation, and learning.

Xu et al. (2012). This paper is one of many that describe SATzilla, a successful implementation of metareasoning. In particular, SATzilla uses algorithm selection to solve the propositional satisfiability problem (SAT). This is not directly applicable to robots, but it demonstrates the benefits of metareasoning and the power of using "niche" algorithms that are superior in a limited set of situations.

Svegliato and Zilberstein (2018). In this paper, Svegliato and Zilberstein proposed an adaptive metareasoning approach based on reinforcement learning that does not need extensive offline work to characterize performance profiles.

Parashar and Goel (2021). Parashar and Goel presented a framework for using metareasoning to learn from failure. They applied this to an assembly robot that uses multi-layer hierarchical assembly planning. The metareasoning approach has multiple structures that interact with different reasoning processes in these layers. This example demonstrates that metareasoning can be implemented in different ways.

1.4   Metareasoning Engineering

Now that we understand what metareasoning is and why we might want to use it on our robot, we should consider how we're going to use metareasoning.

Developing a metareasoning approach for a robot should be a thoughtful process that follows good engineering practice. As a part of the robot's reasoning system, the metareasoning approach should be designed, implemented, and tested along with the rest of the robot. Murphy (2019) presented a systematic approach for designing an autonomous robot; this process emphasizes that a robot should be designed to perform a specific task in a specific environment. (Attempting to design a robot that can do "everything" is doomed to fail.) We agree, which is why this book is not concerned with general purpose metareasoning or theoretical developments related to metareasoning. Instead, we're focused on developing a metareasoning approach that can improve a specific aspect of a robot's reasoning process.

This section briefly discusses the metareasoning engineering process (cf. Fig. 1.3). The discussion here focuses on developing the metareasoning approach, but that should be done in the context of developing the robot that will employ the metareasoning approach; it is not an independent effort.

Metareasoning design. The first step is to identify the object-level reasoning processes that need to be monitored and controlled by the meta-level. This leads to considering the metareasoning design options, which are discussed in Chap. 2.

Architecture. The next step is the metareasoning architecture, which specifies how metareasoning will be implemented in the robot and as part of the larger autonomy software architecture. Chapter 3 discusses the architecture.

Synthesis. The third step synthesizes (creates) the metareasoning policy, which specifies the logic and computations that the meta-level will use to control the object-level

Fig. 1.3 The metareasoning
engineering process has four
steps

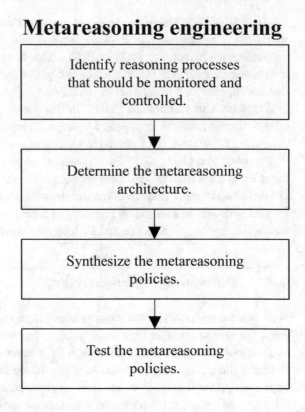

Metareasoning engineering

Identify reasoning processes
that should be monitored and
controlled.

Determine the metareasoning
architecture.

Synthesize the metareasoning
policies.

Test the metareasoning
policies.

reasoning process. If the meta-level should control multiple reasoning processes, then it will need multiple metareasoning policies. Chapter 4 discusses the synthesis process.

Testing and evaluation. The last step tests the metareasoning policy to evaluate its performance, which might lead to accepting the policy as-is, updating the policy, or deciding to scrap it. Chapter 5 discusses approaches for testing metareasoning policies.

Maximizing performance is an important motivation for using metareasoning. Ideally, the metareasoning approach implemented yields a *bounded optimal* robot that behaves as well as possible subject to the limitations of its finite computational resources (Russell and Subramanian 1994). Because this can be difficult to achieve, however, this book is concerned with the practical question of whether adding metareasoning improves the robot's performance.

1.5   Is It Metareasoning or Not?

A robot's meta-level is a computational process controlling another computational process. Moreover, it is sometimes useful to implement a metareasoning policy within the code that is running the object level reasoning processes. Because of these characteristics, the boundary between metareasoning and object-level reasoning can sometimes be hard to identify. My colleagues and I have had spirited arguments sometimes about whether a certain algorithm is metareasoning or not.

Some object-level reasoning algorithms are very sophisticated, and they seem to be metareasoning. For example, consider a safety watchdog algorithm that monitors the commands sent by an unmanned vehicle's motion planning process. If those commands would direct the vehicle into an area that is off-limits, the watchdog overrides those commands and creates a new plan that keeps the vehicle from the danger zone.

Is the watchdog metareasoning or not? On the one hand, it seems to be modifying the object-level reasoning process (by overriding the motion planning process). On the other hand, however, we would argue that it is not metareasoning because it is merely part of the object level reasoning process, which includes this safety check as part of its computations and replaces the dangerous commands with new ones. It is not monitoring and controlling the reasoning process; instead, it is monitoring and modifying the solution generated by the reasoning process.

A competence-aware system (CAS) is another interesting case. A CAS has a human supervisor who can intervene to prevent unsafe actions, but it also has multiple levels of autonomy (Basich et al. 2023). Its planning procedure can select not only an action to perform but also the level of autonomy, which determines what the CAS will ask its supervisor to do (which can range from nothing to manual operation). This is not monitoring and controlling a reasoning process; it is determining how the action will be executed. On the other hand, if we focus on the process of checking the safety of the selected action, which is a reasoning process, then choosing the level of autonomy (i.e., what the human supervisor will do) controls how that reasoning process is done. For instance, the CAS can check the safety itself or ask the human to check it (which is equivalent to delegating the safety check to another agent). That resembles metareasoning.

We would also distinguish between (a) an object level algorithm's failure to find any solution at all and (b) the same algorithm yielding a poor quality solution. In case (a), there is no solution, and the robot will be unable to continue its mission, so we might want a metareasoning approach that reacts to that computational failure (perhaps by invoking another algorithm). In case (b), however, there is a solution, and the robot can follow that, even if it is costly, time-consuming, or dangerous, so we might want another object level algorithm that can improve or fix the solution somehow. This object level algorithm is not metareasoning, however.

Finally, some approaches to computational introspection effectively monitor the object level but do not control the reasoning process. For example, Rabiee and Biswas (2019) described an introspection model that monitored a robot's stereo vision perception pipeline and predicted which parts of the image were likely to confuse the perception process (a reasoning process in the robot's object level). Glare, for instance, might cause the perception process to detect an obstacle that is not truly there. Rabiee and Biswas suggested that their introspection model can help engineers debug an obstacle detection algorithm, but their paper did not describe a metareasoning approach that controls the perception process. Thus, this introspection model, by itself, is not metareasoning, although one might implement a metareasoning approach that uses the predictions from the introspection model to select a different obstacle detection algorithm that was less likely to make mistakes.

Our advice for making this distinction is to consider the function first: an algorithm is performing metareasoning if and only if it is monitoring and controlling a reasoning process. Or consider whether removing the algorithm would disable the robot's object-level reasoning. A metareasoning algorithm is not an essential part of the reasoning process. If removing it disables the reasoning process, then the algorithm is performing reasoning (not metareasoning). If the object-level reasoning still works (perhaps poorly) without this algorithm, then it is metareasoning.

We hope that the material and examples in this book will help us make this distinction correctly, but there are likely more subtle "edge cases" that are closer to the boundary between metareasoning and object-level reasoning, and reasonable people might disagree about whether the procedure in question is metareasoning or not.

1.6   Summary

This chapter has introduced key concepts related to metareasoning and explained why metareasoning should be considered when designing a robot (because it can improve performance and make it safer and more reliable). It listed some key articles and books that provide more information about metareasoning, and it described the process of developing a metareasoning approach.

Metareasoning might seem like a luxury. A robot doesn't need metareasoning to accomplish its mission, and many authors have nothing to say about metareasoning. It might be tempting to ignore metareasoning and skip the additional time and effort that developing a metareasoning approach will require.

Moreover, using a metareasoning approach to defer design decisions (as mentioned in Sect. 1.2) means adding a meta-level, which increases the size and complexity of the robot's software. The benefits of metareasoning might be limited to a small set of scenarios and tasks, and the overhead for metareasoning might degrade performance.

This is a book about metareasoning, so naturally we're optimistic about using metareasoning to improve a robot's performance and convinced that the benefit is likely to be worth the investment of time and effort. Still, like most research and development processes, implementing a new procedure can be time-consuming, tedious, and occasionally frustrating. Perhaps metareasoning is more valuable to those who have already developed a robot and are beginning to design a new one; they're ready to consider new approaches that will lead to better performance and are experienced enough that the additional work seems reasonable.

In any case, the remaining chapters in this book will cover the steps of developing a metareasoning approach. By going through these steps, we'll cover many important decisions that need to be made. The material in these chapters should explain the process sufficiently well that the way forward is clear. Chapter 2 presents many examples of successful metareasoning approaches, which should inspire and encourage those who are working to use metareasoning. Good luck!

References

Ackerman, R., Thompson, V.A.: Meta-reasoning: monitoring and control of thinking and reasoning. Trends Cogn. Sci. **21**(8), 607–617 (2017)

Alterovitz, R., Koenig, S., Likhachev, M.: Robot planning in the real world: research challenges and opportunities. AI Mag. **37**(2), 76–84 (2016)

Anderson, M.L., Oates, T.: A review of recent research in metareasoning and metalearning. AI Mag. **28**(1), 7–16 (2007)

Basich, C., Svegliato, J., Wray, K.H., Witwicki, S., Biswas, J., Zilberstein, S.: Competence-aware systems. Artif. Intell. **316**, 103844 (2023)

Boddy, M., Dean, T.L.: Deliberation scheduling for problem solving in time-constrained environments. Artif. Intell. **67**(2), 245–285 (1994)

Boeing: X-37B. https://www.boeing.com/defense/autonomous-systems/x37b/index.page (2022). Accessed 10 Nov 2022

Brachman, R.J.: Systems that know what they're doing. IEEE Intell. Syst. **17**(6), 67–71 (2002)

Carrillo, E., Jaffar, M.K.M., Nayak, S., Patel, R., Yeotikar, S., Azarm, S., Herrmann, J.W., Otte, M., Xu, H.: Communication-aware multi-agent metareasoning for decentralized task allocation. IEEE Access **9**, 98712–98730 (2021)

Costantini, S.: Meta-reasoning: a survey. In: Kakas, A.C., Sadri, F. (eds.) Computational Logic: Logic Programming and Beyond, pp. 253–288. Springer, Berlin (2002)

Cox, M.T.: Metacognition in computation: a selected research review. Artif. Intell. **169**(2), 104–141 (2005)

Cox, M.T., Raja, A. (eds.): The MIT Press, Cambridge, Massachusetts (2011)

Garey, M.R., Johnson, D.S.: Computers and Intractability: A Guide to the Theory of NP-Completeness. W.H. Freeman and Company, New York (1979)

Genesereth, M.R., Nilsson, N.J.: Logical Foundations of Artificial Intelligence. Morgan Kaufmann Publishers, Palo Alto, California (1987)

Gigerenzer, G., Selten, R.: Rethinking rationality. In: Gigerenzer, G., Selten, R. (eds.) Bounded Rationality: The Adaptive Toolbox, pp. 1–12. The MIT Press, Cambridge, Massachusetts (2001)

Goel, A.K., Fitzgerald, T., Parashar, P.: Analogy and metareasoning: cognitive strategies for robot learning. In: Lawless, W.F., Mittu, R., Sofge, D.A. (eds.) Human-Machine Shared Contexts, pp. 23–44. Academic Press, London (2020)

Good, I.J.: Good Thinking. University of Minnesota Press, Minneapolis (1983)

Greenwald, L.: Analysis and design of on-line decision-making solutions for time-critical planning and scheduling under uncertainty. Dissertation, Brown University (1996)

Horvitz, E.J.: Reasoning about beliefs and actions under computational resource constraints. In: Proceedings of the Third AAAI Workshop on Uncertainty in Artificial Intelligence, pp. 429–444, Seattle, Washington (1987)

Horvitz, E.: Principles and applications of continual computation. Artif. Intell. **126**(1–2), 159–196 (2001)

ISO: ISO 8373:2021 Robotics—Vocabulary. https://www.iso.org/obp/ui/#iso:std:iso:8373:ed-3: v1:en (2021). Accessed 22 Feb 2022

Molnar, S.L., Mueller, M., MacPherson, R., Rhoads, L., Herrmann, J.W.: Metareasoning to improve global and local path planning for a mobile ground robot. Technical Report, Institute for Systems Research, University of Maryland, College Park. http://hdl.handle.net/1903/29723 (2023)

Murphy, R.R.: Introduction to AI Robotics, 2nd edn. The MIT Press, Cambridge, Massachusetts (2019)

Office of the Chief Scientist: Autonomous Horizons: System Autonomy in the Air Force–A Path to the Future, Volume I: Human Autonomy Teaming, United States Air Force AF/ST TR 15-01 (2015)

Paul, J., Stechele, W., Kröhnert. M., Asfour, T.: Resource-aware programming for robotic vision. arXiv:1405.2908 (2014)

Parashar, P., Goel, A.K.: Meta-reasoning in assembly robots. In: Systems Engineering and Artificial Intelligence, pp. 425–449. Springer, Cham (2021)

Rabiee, S., Biswas, J.: IVOA: introspective vision for obstacle avoidance. In: 2019 IEEE/RSJ International Conference on Intelligent Robots and Systems (IROS), pp. 1230–1235 (2019)

Russell, S.J., Subramanian, D.: Provably bounded-optimal agents. J. Artif. Intell. Res. **2**, 575–609 (1994)

Russell, S., Wefald, E.: Do the Right Thing. The MIT Press, Cambridge, Massachusetts (1991a)

Russell, S., Wefald, E.: Principles of metareasoning. Artif. Intell. **49**(1–3), 361–395 (1991b)

Siegwart, R., Nourbakhsh, I.R., Scaramuzza, D.: Introduction to Autonomous Mobile Robots, 2nd edn. The MIT Press, Cambridge, Massachusetts (2011)

Simon, H.A.: The Sciences of the Artificial, 2nd edn. The MIT Press, Cambridge, Massachusetts (1981)

Svegliato, J., Zilberstein, S.: Adaptive metareasoning for bounded rational agents. In: CAI-ECAI Workshop on Architectures and Evaluation for Generality, Autonomy and Progress in AI (AEGAP). Stockholm, Sweden (2018)

Svegliato, J., Wray, K.H., Witwicki, S.J., Biswas, J., Zilberstein, S.: Belief space metareasoning for exception recovery. In: 2019 IEEE/RSJ International Conference on Intelligent Robots and Systems (IROS), pp. 1224–1229 (2019)

Xu, L., Hutter, F., Shen, J., Hoos, H.H., Leyton-Brown, K.: SATzilla2012: improved algorithm selection based on cost-sensitive classification models. In: Proceedings of SAT Challenge, pp. 57–58 (2012)

Zilberstein, S.: Operational rationality through compilation of anytime algorithms. Dissertation, University of California, Berkeley (1993)

Zilberstein, S., Russell, S.: Approximate reasoning using anytime algorithms. In: Natarajan, S. (ed.) Imprecise and Approximate Computation, pp. 43–62. Springer, Boston, Massachusetts (1995)

Metareasoning Design Options

We start the process of developing a metareasoning approach by considering the different options that are available for metareasoning approaches and policies. To do that, this chapter presents examples from our own work and the literature to illustrate the techniques, and it discusses the advantages (and drawbacks) of different options.

By making these decisions about the metareasoning approach, we are beginning to specify how the robot will perform metareasoning. As shown in Fig. 2.1, this is the first step in the metareasoning engineering process that was discussed in Chap. 1. If we find, when we test the approach, that these choices do not work as expected, then we can iterate and revisit these decisions.

Although this chapter describes many ways that a robot can use metareasoning, it is not meant to be a comprehensive catalog or morphological chart. We can use the approaches and examples mentioned here as possible solutions and as inspirations in the search for novel metareasoning approaches that are specific to the robot that we're developing.

2.1 Controlling Reasoning

Because a robot's object level can have many reasoning processes, there might be many options for which object-level reasoning process (or processes) the meta-level should control. Typical object-level reasoning processes include the perception, navigation, cartographer, planning, and motor schema subsystems (Murphy 2019). These subsystems will contain one or more reasoning processes that the meta-level can monitor and control.

Thus, we are faced with the following question: *Which reasoning processes should the meta-level control?*

We can consider different ways to answer this. We might consider which computational resources are the most utilized and which reasoning processes are consuming those

© The Author(s), under exclusive license to Springer Nature Switzerland AG 2023
J. W. Herrmann, *Metareasoning for Robots*, Synthesis Lectures on Computer Science,
https://doi.org/10.1007/978-3-031-32237-2_2

Fig. 2.1 The metareasoning engineering process has steps that lead to one or more metareasoning policies and evidence about their performance. Although shown as a sequence of steps, it is often necessary to return to a previous step to repair issues that arise

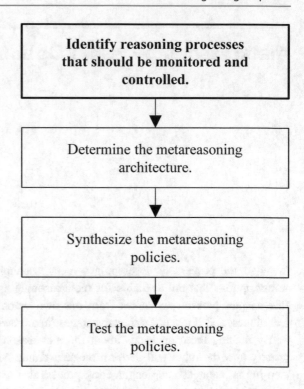

resources. We can use bottleneck analysis to generate such insights, if needed. If the robot's computer is a Linux machine, then we can use the "top" command to see the CPU and memory usage of the programs that are running on the computer. Figure 2.2 shows the output of running the "top" command on the computer that is onboard our Clearpath Jackal robot. In addition, we might consider the reasoning processes that are most likely to fail.

Slowing, halting, or using substitutes for the reasoning processes that consume the most computational resources can be candidate metareasoning modes (see Sect. 2.2). Controlling these reasoning processes should free resources for other reasoning processes, which might lead to better robot performance. Highly utilized CPUs and GPUs will consume more power and experience temperature increases, which can affect their performance and shorten their expected life; thus, reducing their utilization is desirable.

Identifying poorly performing reasoning processes can lead to candidate metareasoning modes. If a reasoning process sometimes yields poor solutions (such as unsafe actions, excessively long routes or paths, or very inaccurate results) or fails to generate any solution, then modifying that process (by switching to another algorithm, for instance) might yield better performance. Metareasoning has been used to help a robot recover from reasoning failures; see, for instance, Molnar et al. (2023), Parashar et al. (2018) and Svegliato et al. (2019). Rabiee and Biswas (2020) used metareasoning to prevent mistakes by a visual simultaneous localization and mapping (VSLAM) process.

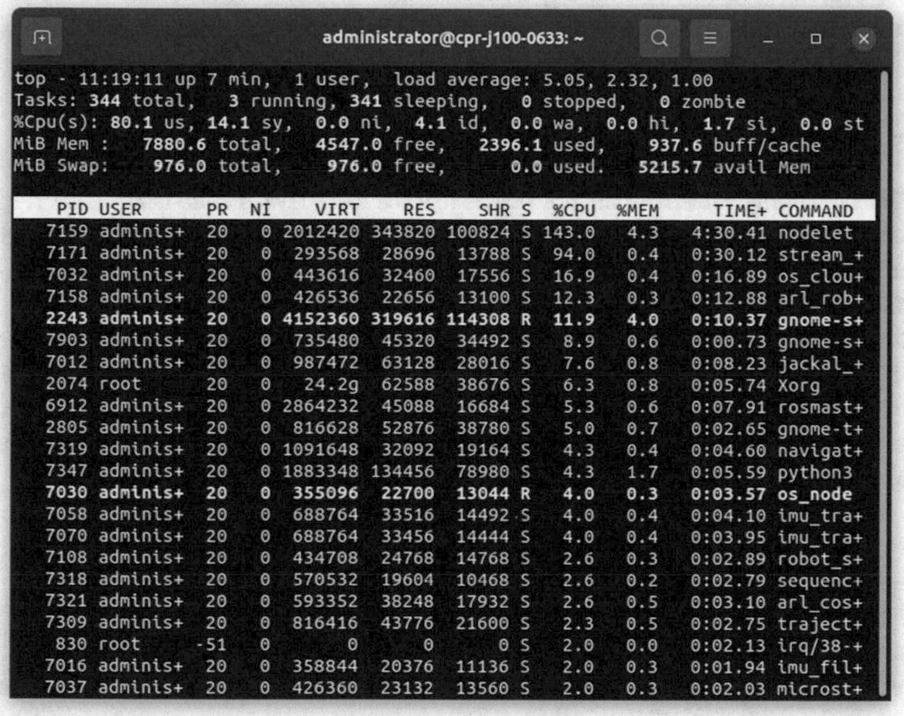

```
                        administrator@cpr-j100-0633: ~              Q  ≡   _  □  ×

top - 11:19:11 up 7 min,  1 user,  load average: 5.05, 2.32, 1.00
Tasks: 344 total,   3 running, 341 sleeping,   0 stopped,   0 zombie
%Cpu(s): 80.1 us, 14.1 sy,  0.0 ni,  4.1 id,  0.0 wa,  0.0 hi,  1.7 si,  0.0 st
MiB Mem :   7880.6 total,   4547.0 free,   2396.1 used,    937.6 buff/cache
MiB Swap:    976.0 total,    976.0 free,      0.0 used.   5215.7 avail Mem

   PID USER       PR  NI    VIRT    RES    SHR S  %CPU  %MEM     TIME+ COMMAND
  7159 adminis+   20   0 2012420 343820 100824 S 143.0   4.3   4:30.41 nodelet
  7171 adminis+   20   0  293568  28696  13788 S  94.0   0.4   0:30.12 stream_+
  7032 adminis+   20   0  443616  32460  17556 S  16.9   0.4   0:16.89 os_clou+
  7158 adminis+   20   0  426536  22656  13100 S  12.3   0.3   0:12.88 arl_rob+
  2243 adminis+   20   0 4152360 319616 114308 R  11.9   4.0   0:10.37 gnome-s+
  7903 adminis+   20   0  735480  45320  34492 S   8.9   0.6   0:00.73 gnome-s+
  7012 adminis+   20   0  987472  63128  28016 S   7.6   0.8   0:08.23 jackal_+
  2074 root       20   0   24.2g  62588  38676 S   6.3   0.8   0:05.74 Xorg
  6912 adminis+   20   0 2864232  45088  16684 S   5.3   0.6   0:07.91 rosmast+
  2805 adminis+   20   0  816628  52876  38780 S   5.0   0.7   0:02.65 gnome-t+
  7319 adminis+   20   0 1091648  32092  19164 S   4.3   0.4   0:04.60 navigat+
  7347 adminis+   20   0 1883348 134456  78980 S   4.3   1.7   0:05.59 python3
  7030 adminis+   20   0  355096  22700  13044 R   4.0   0.3   0:03.57 os_node
  7058 adminis+   20   0  688764  33516  14492 S   4.0   0.4   0:04.10 imu_tra+
  7070 adminis+   20   0  688764  33456  14444 S   4.0   0.4   0:03.95 imu_tra+
  7108 adminis+   20   0  434708  24768  14768 S   2.6   0.3   0:02.89 robot_s+
  7318 adminis+   20   0  570532  19604  10468 S   2.6   0.2   0:02.79 sequenc+
  7321 adminis+   20   0  593352  38248  17932 S   2.6   0.5   0:03.10 arl_cos+
  7309 adminis+   20   0  816416  43776  21600 S   2.3   0.5   0:02.75 traject+
   830 root      -51   0       0      0      0 S   2.0   0.0   0:02.13 irq/38-+
  7016 adminis+   20   0  358844  20376  11136 S   2.0   0.3   0:01.94 imu_fil+
  7037 adminis+   20   0  426360  23132  13560 S   2.0   0.3   0:02.03 microst+
```

Fig. 2.2 The output of a top command on a Clearpath Jackal (*Image* Kevin Carey)

In an autonomous robot, groups of object-level activities are arranged using an architecture, and many architectures have been developed (Arkin 1998). A hybrid deliberative/reactive operational architecture has three layers: reactive, deliberative, and interactive (Murphy 2019); note that all three layers are part of the robot's object level. If a robot's reactive layer has tightly coupled behaviors that include sensing and acting, then it might be impractical to incorporate metareasoning there because the metareasoning process might cause undesirable delays. A robot's deliberative layer, which usually includes planning activities, is not as fast as its reactive layer and is thus a more natural focus for metareasoning. The robot's interactive layer communicates with other robots and with humans, and metareasoning can be used to control these activities.

In principle, we should consider controlling all of the object-level reasoning processes with a comprehensive metareasoning approach. This will likely be very complicated, however, and whether the additional benefits of a comprehensive metareasoning approach are worth the effort is an open question. The Pareto principle implies that we might get most of the benefit by controlling a smaller number of reasoning processes.

Therefore, for now, we'll identify a small number of reasoning processes for the meta-level to control. The next step is to consider what the meta-level will do.

2.2   Metareasoning Modes

After identifying some reasoning processes that the meta-level should control, we'll consider the metareasoning mode. A meta-level controls object-level reasoning by making decisions about what a reasoning process should do. The meta-level uses a metareasoning policy to make that decision, and the result of that decision is the metareasoning mode. The metareasoning mode describes what is specified or modified as the result of metareasoning. The metareasoning mode is closely related to the decision that the meta-level is making and the instructions that the meta-level sends to the object level. For instance, if the meta-level determines when an anytime reasoning algorithm should halt (and produce its output), then the metareasoning mode is the stopping time.

The metareasoning mode describes "what" the meta-level produces; it does not describe "how" the meta-level produces it (i.e., how the meta-level decides). The metareasoning policy describes "how."

The following paragraphs include options for metareasoning modes, along with references to works that give details and examples. For reference, Table 2.1 briefly describes these modes. Some of these are more specific, while others are more general, and some overlap. Langlois et al. (2020) provided additional examples of metareasoning modes for

Table 2.1 Metareasoning modes

Metareasoning mode	The meta-level decision
Algorithm selection	Which algorithm should be executed?
Parameter adjustment	What value should the parameter have?
Policy adjustment	How should be the policy be modified?
Stopping an algorithm	For how long should the algorithm run?
Controlling an algorithm	What should the algorithm do next?
Pausing an algorithm	When and for how long should the algorithm pause?
Pausing communication	When and for how long should the robot stop communicating?
Offloading computation	When should another resource perform reasoning?

multi-robot systems. Creative metareasoning approaches that use modes beyond the ones listed here are possible as well.

Algorithm selection. The meta-level selects the algorithm that the reasoning process should execute to carry out its function. Typically, the meta-level has a predetermined set of candidate algorithms, and it chooses one of those. Chapter 1 reviewed the key works on algorithm selection, which is an important metareasoning problem. It can be used in many settings. For computer vision applications, Lee et al. (2021) and Dawson and Herrmann (2022) described metareasoning approaches that selected an object detector. In a simulated multi-agent system that was performing a search mission, the agents used algorithm selection to determine which collaboration algorithm they should use to allocate tasks among themselves (Carrillo et al. 2021). Song et al. (2022) added a meta-level to a biped robot's control system; this meta-level used a partition-based metareasoning policy that selected a control strategy to stabilize the robot in response to a push. Based on the position and velocity of the robot's center of mass, the metareasoning policy chooses the ankle strategy, the hip strategy, or the step strategy. Each strategy is a different controller that attempts to stabilize the robot.

An autonomous mobile robot uses path planning to determine the best way to get to a goal location. In practice, a path planning method might fail because it cannot find a feasible solution, which might be due to a problem with the algorithm (or its implementation) or an issue with the graph representation that it is using. If the robot has no way to recover from this failure, then it will be unable to continue, and it cannot complete its mission. Molnar et al. (2023) developed multiple metareasoning approaches for switching path planners when the current active planner cannot find a feasible path. They implemented the approaches within a ROS-based autonomy stack and conducted simulation experiments to evaluate their performance in multiple scenarios. Their results show that these metareasoning approaches reduce the frequency of failures and reduce the time required to complete the mission.

Parameter adjustment. The meta-level modifies the value of one or more parameters (sometimes called "knobs") that affect the reasoning process or the algorithms that the reasoning process executes. If the parameter value directly or indirectly affects the time required to execute the reasoning process, then this mode is related to controlling anytime algorithms and determining when to stop an algorithm (discussed below). More generally, adjusting the parameter value influences the time required, the computational resources required, and the quality of the output. The following paragraphs describe some recent examples.

For controlling computer vision on a mobile device, Lee et al. (2021) presented a metareasoning approach that dynamically changes values for key algorithm parameters: (1) the resolution of the input image fed into the detector, (2) the number of object candidates (proposals) that the first stage returns, (3) the specific combination of the four feature maps used in the object detector, (4) the resizing factor for the object tracker's input image, (5) the number of objects to track, and (6) the number of frames processed by the object tracker between runs of the object detector.

For an unmanned aerial vehicle (UAV) that needs to fly quickly through a forest, Jarin-Lipschitz et al. (2022) described a planning approach that used metareasoning to adjust dispersion, a key planning parameter. The object level had three reasoning modules: (1) state estimation and mapping, (2) planning, and (3) control. The planning module uses a minimum-dispersion-based motion primitive planner to determine the trajectories that vehicle should follow. This includes a global planner that constructs a path to the destination and a local minimum-dispersion planner that runs more frequently than the global planner and must meet a time threshold. The value of the dispersion parameter affects the graph and the quality of the path that the local planner creates. A smaller dispersion value yields a graph with more nodes and a better (shorter) path, but this increases the computational effort needed. Changing the dispersion allows one to make tradeoffs between computation time and plan quality. The meta-level used three rules that changed the dispersion value: (1) if the planning time (with the current dispersion) exceeds the time threshold, then increase the dispersion; (2) if the planner fails to find a feasible plan, then decrease the dispersion; and (3) if the planner has been successful and the planning is below the time threshold, then decrease the dispersion. To implement this metareasoning approach, the vehicle was provided with a family of motion planning graphs created with different dispersion values.

Policy adjustment. In some cases, the object-level reasoning algorithm uses a policy to determine which action the robot should perform next; this approach is common if the planning problem is formulated as a Markov decision problem (MDP). A typical MDP is the stochastic shortest path problem (Bertsekas and Tsitsiklis 1996). Lin et al. (2015) proposed metareasoning approaches that, at each time step, either improve the current policy or let the object-level planning continue without changing it. Improving the current policy incurs a cost but might lead to better performance in the future. Parashar et al. (2018) developed a metareasoning approach that reacts to discrepancies between actual and expected state transitions by calling a reinforcement learning (RL) module to update the policy that the robot's planner uses.

Stopping an algorithm. The meta-level determines when to stop the anytime algorithm that the object level is executing. There are two problems that should be distinguished (Zilberstein 1993): If the anytime algorithm is a *contract algorithm*, then the meta-level should determine how long the algorithm should run before it begins (this is known as "passive monitoring"). If the algorithm is an *interruptible anytime algorithm*, then the meta-level should monitor the algorithm as it runs and determine when to stop it (this is known as "active monitoring"). If a robot is using multiple anytime algorithms, then the meta-level must determine how much time to allocate to each algorithm, which is the anytime algorithm composition problem (Zilberstein and Russell 1995).

Many different types of algorithms can be implemented as anytime algorithms, including numerical approximation, search algorithms, approximate dynamic programming,

Monte Carlo algorithms, approximate evaluation of Bayesian belief networks, motion planning, and object detection (Boddy and Dean 1994; Svegliato et al. 2020).

Sung et al. (2021) studied anytime motion planners and proposed different approaches for learning metareasoning policies. The meta-level addressed the problem of when to stop trying to improve a feasible path. They trained a metareasoning policy from a dataset of performance profiles that they generated by running their anytime motion planner on a set of problem instances.

As an example of active monitoring, consider the maintenance robot discussed by Zilberstein (1993). When the robot begins examining a failed machine, it starts at the level of large subsystems and then proceeds down to smaller subsystems (which have fewer components). Given enough time, the robot will eventually identify a specific failed component. At any point in time, the robot has identified some defective subsystem (the one that contains the failed component) and knows the time and cost of replacing it and the cost associated with the unavailable machine. The robot's meta-level must decide whether to stop deliberation (which will lead to an earlier repair) or continue the examination (which will lead to a more specific diagnosis).

Controlling an algorithm. The meta-level modifies the object-level algorithm by telling it what to do next based on the current state of the algorithm, computation time concerns, and other factors. For example, Karpas et al. (2018) used metareasoning to control the Lazy A* algorithm and reduce the time needed to find an optimal solution to shortest path and similar planning problems; in particular, the metareasoning policy determined when the algorithm should use a second heuristic to evaluate a node in the graph. The metareasoning policy uses the time required to compute the second heuristic and the time required to expand the node and evaluate all of its successors to make this decision.

Rabiee and Biswas (2020) used metareasoning to control visual SLAM. They developed an introspection function that uses a convolutional neural network to predict the "reliability" of each part of the input image. The metareasoning policy then tells the SLAM algorithm to sample more features from the reliable parts of the image.

Pausing an algorithm. The meta-level interrupts and suspends the normal reasoning process. The pause can be in the middle of the reasoning process or between distinct runs of the reasoning process. For example, to reduce CPU utilization, the meta-level might insert a short pause between the processing of images from a camera (Dawson 2022; Dawson and Herrmann 2022). The pause will reduce the image processing throughput and the computational resources required. After the reasoning process is paused, the meta-level might determine the length of the pause at the beginning of the pause. If it determines the pause length, then it tells the object level to resume the reasoning process at the end of the pause. If it does not, then the reasoning process remains suspended until the meta-level decides to resume it. (These options correspond to the ideas of passive and active monitoring of anytime algorithms.)

If an unexpected problem has occurred, the meta-level can modify the reasoning process by interrupting its normal execution cycle and commanding that the object level run a different decision-making process. For example, Svegliato (2022) presented a metareasoning approach that responds to exceptions (unanticipated scenarios that the execution process cannot resolve) by pausing the normal decision-making process and invoking an appropriate exception handler. When implemented on an autonomous vehicle, this metareasoning approach successfully handled exceptions (different types of obstacles) during simulations and a road test; a vehicle without metareasoning was unable to complete the route due to the obstacles. (See also Svegliato et al. 2019.)

Pausing communication. In a multi-robot system, a robot's meta-level determines that the robots should not communicate for a while. Like pausing an algorithm (discussed earlier), the meta-level might determine the length of the pause at the beginning of the pause (or not). Suspending communication will affect reasoning processes that expect inputs from other robots. Without these inputs, these reasoning processes will either reuse the previously received inputs or estimate what the inputs should be. Although a pause should reduce the resource consumption of the communication processes, using outdated inputs might lead to poor solutions from the affected reasoning processes.

Offloading computation. In some cases, a robot can offload (or delegate) a reasoning process to another robot or computer with which it can communicate. The metareasoning policy decides when to do this offloading. For example, Dasari et al. (2019) developed an offloading approach for video stream processing. The robot's meta-level uses a classifier to determine the computational complexity of the processing task effort required and then decides which device or computer should process the incoming video.

The metareasoning approach that Navardi and Mohsenin (2023) implemented to control the navigation process on a small UAV also used this mode. Their UAV (a Crazyflie) had the ability to use an on-board (edge) model or a cloud-based model. The latter model was more accurate, but the latency depended upon the distance; thus, as the UAV moved farther away from the server, the latency increased to unacceptable levels. The metareasoning policy determined when to offload the navigation task to the server, which would use the more accurate cloud-based model, and when to perform that reasoning task onboard, using the edge model. (Note that this also was a type of algorithm selection because the two models were different.)

This review of metareasoning modes that have been used in previous work shows the variety of ways that metareasoning can control object-level reasoning. The meta-level can select an algorithm, set its parameters, control its operations, determine when to stop it, or delegate it to another resource. One might invent other metareasoning modes that are not listed here. In some cases, the best choice of metareasoning mode might follow immediately from the computational problems that the robot experiences. In other cases, we might need to implement and test a few different metareasoning modes to determine which (if any) will improve the robot's performance (Chapter 5 discusses testing in detail).

2.3   Policy Options

In principle, metareasoning should consider all of the computational options and select the "best" one. Much previous research on metareasoning has proposed how to formulate the metareasoning problem (selecting the "best" computational option) and how to solve it (cf. Boddy and Dean 1994; Etzioni 1989; Russell and Wefald 1991a, b; Zilberstein 1993).

Generally, the meta-level determines when the expected benefit of additional computation outweighs the cost of the delay that additional computation will create. For example, if the reasoning algorithm is a search algorithm, continuing the search might yield a solution that is better than the best one found so far, which would be a benefit, but continuing the search delays the time until the object level can process that solution and tell the robot what it should do next.

Performing one computational action requires not performing others, which leads to an opportunity cost. The problem of computing the expected opportunity cost of a computational action is NP-complete, however (Etzioni 1989). An NP-complete problem is inherently intractable, so no algorithm can solve the problem in a reasonable amount of time. The theory of NP-completeness provides ways to prove that a complex problem is inherently intractable (Garey and Johnson 1979). Other metareasoning problems are also difficult; for instance, the general problem of compilation of composite expressions is NP-complete in the strong sense (Zilberstein 1993). Lin et al. (2015) found that metareasoning for planning under uncertainty is harder than solving a MDP and concluded that "optimal general metareasoning is impractical."

Because the metareasoning problem is NP-complete in general, it will be difficult to solve this problem in reasonable time. That is, finding the best computation is as difficult as finding the sequence of cities that minimizes the total distance in a traveling salesman problem (TSP). Moreover, even for special cases where an optimal solution can be found quickly, the optimality of the meta-level's solution algorithm will depend exactly on the conditions that make that case special; this solution algorithm will likely generate suboptimal solutions in situations that don't meet those conditions.

Therefore, just as mathematicians and computer scientists have developed algorithms (heuristics) that can quickly generate high-quality (not necessarily optimal) solutions to complex optimization problems, we need to consider metareasoning policies that can make high-quality metareasoning decisions quickly, even if they are not guaranteed to produce optimal decisions.

The following paragraphs describe some options for metareasoning policies. Some of these require models that describe the performance of reasoning processes. These models are generally created "offline" but used "online."

In general, for robotic systems, there are two classes of computations: "offline" and "online." When we mention offline computations, we're referring to computational work that is done before the robot begins operating and starts its mission. Offline computations are used to generate performance profiles or to evaluate thresholds for metareasoning policies.

When we mention online computations, on the other hand, we're referring to computations that the robot does while it is operating and pursuing its mission. A robot's object level performs path planning online, for instance. Online computations in the meta-level might include evaluating the rules that form a metareasoning policy.

Offline computations are not "free," for they require time during the development of a metareasoning approach. But online computations are more "expensive" because they consume the robot's computational resources, which might be quite limited during its operation, and waiting for them to complete might delay the robot's activities.

2.3.1 Optimization

Optimization-based metareasoning approaches formulate a metareasoning decision as an optimization problem and then identify the best object level reasoning alternative.

The metareasoning decision can be formulated using decision theory. A key part of using decision theory is the utility function $U(W)$. This function can assign a real number to any possible state W in the state space (Herrmann 2015). Ideally, the robot acts in a way that leads to states with greater utility, and the objective of any decision is to maximize the utility of the resulting state. In the presence of uncertainty, the objective is to maximize the expected utility.

In principle, the meta-level should select the computational action that maximizes the expected utility of the next state and all future states. Because this can be a difficult problem, a slightly simpler approach is to select the computational action that has the most benefit (the greatest expected utility of the next state) (Russell and Wefald 1991a, b). This requires making various assumptions about the irrelevance of other actions and the future (Boddy and Dean 1994), which is why it can be called a myopic approach.

Using this type of approach requires considering the intrinsic value of a possible reasoning process; this value is independent of the computational time (and other resources) required to perform the process. It is usually based on the "quality" of its output. The expected value of an anytime algorithm increases if it is given more time; that is, the expected quality of its output increases over time. A performance profile describes the expected intrinsic value as a function of time (Boddy and Dean 1994). Figure 2.3 shows a performance profile derived from experimental results of a search algorithm that constructs solutions to a TSP. The relative solution deviation measures the difference between the length of the solution and the length of the best solution found; thus, the algorithm's value increases as this deviation decreases. Naturally, the actual value of the solution that is found can vary, so a performance profile is not a complete description of the reasoning process's performance.

Fig. 2.3 A performance profile for a search algorithm that finds solutions for traveling salesman problems

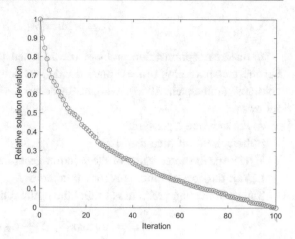

On the other hand, while it is waiting for the reasoning process, the robot is not accomplishing its mission. If it is moving or acting, it might be doing something counterproductive. Consider, for example, a robot that continues moving while it waits for its planning algorithm to determine a new path to the goal. The robot might be moving in the wrong direction; if so, that unfortunate motion will further increase the time needed to get to the goal. Thus, giving the reasoning process more time is undesirable, and the total value of a reasoning process combines its intrinsic value and the "cost" of the time required to perform it.

For example, the following discrete metareasoning problem is a simplification of one presented by Russell and Wefald (1991a, b). First, let us define the following notation:

α_0: the default external action that the agent will perform if no further computation is done;

$S = \{S_1, \ldots, S_n\}$: the set of possible computational actions that the agent can perform;

$\alpha(S)$: the external action that computational action S will recommend.

W: a world state.

$U(W)$: the agent's utility in state W.

$[X]$: the world state that results from taking action X.

$|S|$: the elapsed time required to perform computational action S.

$TC(t)$: time cost due to a delay of t time units.

$V(S)$: the net value of performing computational action S (instead of action α_0).

$$V(S) = U([\alpha(S)]) - U([\alpha_0]) - TC(|S|)$$

Note that the value function includes the benefit of increasing utility (by performing a computational action) and the cost of the delay incurred by performing that computational action.

The optimization problem is to find the optimal computational action S^*:

$$S^* = \text{argmax}\{V(S){:}S \in \mathbf{S}\}$$

A different optimization problem occurs when there is only one anytime algorithm but the metareasoning process must decide how long to let it run. This time is a "fixed contract" (Zilberstein 1993). We can formulate a simplified version of this problem as follows:

A: the anytime algorithm.

t: the amount of time that A runs.

$V_A(t)$: the expected value of the solution generated by running A for t time units.

$C_A(t)$: the cost of running A for t time units.

The optimization problem is to find the optimal time t^*:

$$t^* = \arg\max\{V_A(t) - C_A(t), t > 0\}$$

An important metareasoning problem is the meta-level control problem, which seeks to optimize a utility function that depends upon the quality of the solution that the anytime algorithm creates and the time spent to computing it. This can be solved using a monitoring and control approach that monitors an anytime algorithm in real-time to determine when to stop it. Following Svegliato et al. (2020), the problem can be formulated as a Markov decision process (MDP); a solution to the MDP is a policy π that specifies an action a (either Stop or Continue) for the current state, which is described by two variables: q, the quality of the best solution that the anytime has found so far, and t, the time spent computing. That is, $a = \pi(q, t)$. The meta-level monitors the state variables while the anytime algorithm is running, and the meta-level stops the anytime algorithm when the state reaches a point where the policy $\pi(q, t) = $ Stop.

There are numerous ways to create this policy. In an offline approach, we would run the anytime algorithm on many instances of the problem, which yields performance profiles that describe how the solution quality improves as a function of computation time.

We can also use an online model-free reinforcement learning approach that updates the policy every time that the anytime algorithm is run (Svegliato et al. 2020). This avoids the effort of the offline experiments, and it allows the policy to change if the characteristics of the problem instances change over time. One disadvantage is that the utility will likely be low during the initial runs.

According to Zilberstein (1993), a robot is said to be operationally rational if "it optimizes the allocation of resources to its performance components so as to maximize its overall expected utility in a particular domain." If the variance of the performance profile $V_A(t)$ is small, then using this metareasoning policy will achieve operational rationality (Zilberstein 1993).

Consider the problem that Boddy and Dean (1994) studied. The robot's mission specifies a finite set of locations that the robot should visit; its objective is to minimize the time required to complete this mission. The time required depends upon the route that its object-level route planning algorithm generates. This algorithm, an anytime algorithm, begins with a randomly generated initial route and then searches for better solutions. Whenever it is stopped, it returns the best solution found so far.

The robot cannot begin moving until this algorithm returns a route; thus the total time to complete the mission is the sum of two times: (1) the computation time of the algorithm (until the meta-level stops it) and (2) the time to follow the route. Allowing the route planning algorithm to run longer should yield a shorter route that requires less time. When should the robot's meta-level stop the route planning algorithm?

Suppose that we have a performance profile $f(t)$ that describes the expected quality of the route (the time required to follow the route) as a function of the algorithm's computation time t. (This might resemble the curve shown in Fig. 2.3.) Then, minimizing the expected time requires finding t^*, the optimal computation time:

$$t^* = \arg\min\{t + f(t) : t > 0\}$$

For dynamic control, at time t, let $y(t)$ be the actual quality of the best route that the algorithm has found at time t. (We want to minimize y because it measures the time required to follow the route.) Let Δt be the time step for monitoring the object level. We assume that the improvement (decrease) in y in the next time step will be no more than the improvement in the last time step; that is,

$$y(t) - y(t + \Delta t) \leq y(t - \Delta t) - y(t)$$

Then, the meta-level should stop the route planning algorithm at time t when the value of $y(t)$ is not decreasing sufficiently. That is, when $y(t - \Delta t) - y(t) \leq \Delta t$. At this point,

$$y(t + \Delta t) \geq y(t) - (y(t - \Delta t) - y(t)) \geq y(t) - \Delta t$$

which implies the following:

$$t + \Delta t + y(t + \Delta t) \geq t + y(t)$$

That is, spending another time step searching will not improve the route enough to compensate for the time spent searching, so the total time (computing and traveling) is beginning to increase.

In practice, this policy will stop the search at the first time step in which the search fails to find a better route, which might stop the search too soon. We might instead use a moving average (over the last w time steps) as the stopping policy. Then, the search should stop when the following condition holds:

$$\frac{y(t - w\Delta t) - y(t)}{w} \leq \Delta t$$

If the robot can plan while moving, on the other hand, it might be better for the robot to take a longer, slower route to an intermediate waypoint so that the object level has

more time to plan the rest of the route and generate a better solution; this situation yields
the metareasoning problem that was solved by Cserna et al. (2017).

2.3.2   Deliberation Scheduling

Deliberation scheduling is an optimization-based metareasoning approach that allocates
computational resources based on expectations about how reasoning processes affect system behavior and performance (Boddy and Dean 1994). This scheduling problem involves
allocating computation over a finite time horizon to respond to events that will occur in the
future; the objective is to maximize the total performance of the anytime algorithms that
the object level runs to respond to these events. If the performance profiles are piecewise
linear and have diminishing returns, then this problem can be solved by the "simple deliberation scheduler" that Boddy and Dean described. This approach involves determining
how much time to allocate to each algorithm when there are multiple anytime algorithms
that need to be run; clearly, allocating more computation time to one algorithm requires
allocating less to one or more others.

2.3.3   Heuristics

Metareasoning heuristics use a pre-determined policy to make a metareasoning decision
quickly. A metareasoning policy is a rule (or set of rules) that the meta-level uses to control the object level reasoning. Determining a good metareasoning policy is the purpose
of the synthesis step that Chap. 4 presents. That process, which is performed off-line,
yields a policy that the robot's meta-level can perform online.

  A heuristic can be a set of rules that use a set of values (thresholds) to determine the
metareasoning decision.

  Let $x(t)$ be the value of the state variable at time t. This state variable describes the
single factor that the meta-level monitors for controlling reasoning. (In practice, of course,
there might be many factors that need to be monitored.) Suppose there is a finite set
of possible actions $\{y_1, \ldots, y_n\}$ and a corresponding set of thresholds $\{T_0, T_1, \ldots, T_n\}$.
Then, the metareasoning policy can be a set of rules, one for each $i = 1, \ldots, n$: If
$T_{i-1} < x(t) \leq T_i$, then choose y_i at time t.

  This type of scheme can be extended to cases where there are multiple state variables
and more complicated thresholds. For example, for controlling a biped robot, Song et al.
(2022) used a partition-based metareasoning policy that selects a control strategy based
on the position and velocity of the robot's center of mass (the state variables). The two-
dimensional state space was divided into three regions, and each region corresponds to a
different controller. The metareasoning policy maps the state variables into one of these
regions and then chooses the corresponding controller.

To control the navigation process on a small UAV, Navardi and Mohsenin (2023) implemented a metareasoning approach that considered the latency of two different navigation models and a latency threshold (the time until the UAV would collide with the obstacle in front of it). The cloud-based model was more accurate than the on-board (edge) model, which was optimized for the UAV's limited computational resources. The latency of the cloud-based model depended upon the distance between the server and the drone. The latency of the on-board (edge) model was estimated from offline experiments. The metareasoning policy used a pair of rules: if the cloud-based model's latency is less than the latency threshold, then use the cloud-based model. Otherwise, if the on-board model's latency is less than the latency threshold, then use the on-board model. Finally, if both models' latencies are too large, then slow the UAV in order to increase the latency threshold.

2.3.4 Learning

In the context of robotics, we use the term "learning" in multiple ways. The term refers to the general topic of machine learning (ML), an important branch of artificial intelligence (AI). ML includes many data analysis techniques, including approaches that create and modify a function, such as an artificial neural network, in response to data in order to improve that function. We can employ ML techniques to train a classifier that can detect objects in an image, and we can apply ML techniques to train a regression function that can predict the performance of an algorithm.

A supervised learning approach uses labeled data to train the function. This approach modifies the structure and parameters of the function to minimize the error between the function's output and the true values (from the labeled data). Unsupervised learning approaches such as clustering do not require labeled data; instead, they look for patterns among the items that the data describe. Murphy (2019) discusses learning, ML techniques, and their application to robotics, especially for object level reasoning processes such as object detection and motion planning.

Learning a fixed metareasoning policy. For the purposes of metareasoning, we can use ML techniques, especially supervised learning, to synthesize a metareasoning policy from data about the performance of reasoning options in different situations. For example, if we wanted to synthesize a metareasoning policy for algorithm selection, we can generate a large set of instances, run the different algorithms on all of these instances, and label each instance with the algorithm that performed best on that instance. We can then apply a supervised learning algorithm to generate a classifier that identifies which algorithm should perform best on an input instance. When we implement the metareasoning approach, we include that classifier as the metareasoning policy. The robot's meta-level selects the algorithm that the classifier determines.

Of course, that is not the only way to use ML for metareasoning. Herrmann (2020) used supervised learning to train a regression function that predicts the performance of an algorithm from data about the current environment. This metareasoning policy had multiple regression functions, one for each algorithm option, and it selected the algorithm with the best predicted performance. One benefit of using regression functions (instead of a single classifier) is that, if we change the set of algorithm options, we can easily add (or remove) a regression function, instead of re-training the classifier.

Generally, we would use ML this way when we synthesize the metareasoning policy; Chapter 4 discusses the synthesis process in more detail. The robot benefits from having a good metareasoning policy when it begins operating. On the other hand, the robot cannot change the metareasoning policy.

Learning to metareason. In some cases, however, we might want to give the robot's meta-level the ability to modify its metareasoning policy over time; that is, we want the robot to learn to metareason. This capability will help the robot adapt to slower, long-term changes in the environment.

Although the robot's meta-level can react to changes in the environment and the robot, when we synthesize a metareasoning policy, we must make assumptions about the situations that the robot will encounter. For instance, when developing a metareasoning approach for algorithm selection, we must define the range of the problem instances that the robot's object level needs to consider, and we synthesize the metareasoning policy to improve the robot's performance based on that range.

If the types of situations will change over time in unforeseen ways, however, the original metareasoning policy will become obsolete, and its decisions might degrade the robot's performance (instead of improving it). For instance, perhaps the number of tasks in the robot's mission is increasing over time, which was unanticipated when the robot was developed. The robot's metareasoning policy, which was created to select the best planning algorithm for shorter missions, might select inferior planning algorithms for these more complicated missions.

To reduce the impact of such changes, we might apply reinforcement learning (RL) techniques such as Q-learning. In general, a RL technique tries different actions, observes what happens, and updates a function that estimates the value of performing an action in a given state. This clearly fits the metareasoning paradigm in which the meta-level has information about the state of the environment and the system and selects the reasoning option that has the best value in that state. Giving the meta-level the ability to update its predictions of a reasoning option's value will help it adapt to changes.

Unfortunately, learning a good policy might require many trials, so we would still want to synthesize a good initial metareasoning policy (as described in Chap. 4) so that the robot can perform well when it begins operating. Using RL to update the policy, which should make the meta-level more powerful, is an opportunity for future research and development.

AI researchers have also been developing meta-reinforcement learning approaches so that the agent can leverage prior experience on similar tasks. This type of meta-learning has two steps: the first (meta-training) learns an algorithm; the second (meta-testing) uses that algorithm to learn a policy for the current environment (Caron 2021).

2.4   Structures for Multi-robot Systems

A multi-robot system (MRS) has multiple robots that collaborate to complete a mission. As they perform that mission, they might perform metareasoning to adapt their reasoning processes. The way that they perform metareasoning is their *metareasoning structure*. The metareasoning structure is the relationship (if any) between the robots at the meta-level. This section is based on the survey by Langlois et al. (2020); that paper and the references there have more details and examples.

There are multiple ways to implement metareasoning in a MRS. For example, in some approaches, each robot has its own meta-level that controls its own object-level reasoning. Other approaches use a centralized leader that controls the robots' reasoning. This section describes the different metareasoning structures that previous work has considered.

Independent metareasoning. In the independent metareasoning structure, each robot has a meta-level that performs metareasoning independently of the other robots, although the robots' object levels might communicate. This structure is very common. Because the meta-levels are independent, no additional communication or coordination between the agents is required, so implementing this structure is easier than implementing a decentralized structure. This structure adds the overhead of the meta-level to each agent, which might affect the computational resources available at the object-level. In a cooperative system, moreover, the meta-level should consider how the agent's reasoning affects not only agent-level performance but also system-level performance.

Coupled metareasoning. In the coupled metareasoning structure, each robot has a meta-level that performs metareasoning independently of the other robots except for one caveat: when one meta-level has decided to halt its robot's object-level reasoning, it communicates this decision to the other robots, and those meta-levels halt their object-level reasoning as well. The meta-levels are coupled because one robot's metareasoning decision depends upon another robot's metareasoning decision. In this structure, the interaction goes one way; it is not bidirectional. The meta-levels do not cooperate to make a coordinated decision; the meta-levels work independently but stop simultaneously. In a MRS in which the robots cannot act until they are finished reasoning, this coupling enables the agents to coordinate their behaviors. The communication cost in this structure is lower than the communication cost in the decentralized structure, but the metareasoning decision might be a poor one for robots that needed more time to find a better solution.

Decentralized metareasoning. In the decentralized metareasoning structure, the robots cooperate to determine how they are going to reason. This structure requires more communication and coordination, which might increase the overhead associated with metareasoning, but, by cooperating, the robots might achieve better reasoning.

Multiple metareasoning agents. In this type of structure, the MRS includes additional specialized agents that engage in metareasoning that determine the other robots' reasoning. This structure adds resources (more agents) to perform metareasoning so that the other robots have no additional overhead. Keeping the metareasoning agents informed and transmitting their metareasoning decisions will require more communication.

Centralized metareasoning. In the centralized metareasoning structure, a designated "leader" agent does the metareasoning and tells the other robots how to reason. The leader's objective is to maximize the performance of the entire system. Like the structure with multiple metareasoning agents, this structure adds additional resources but requires additional communication. A single metareasoning agent that has information from the entire MRS should be able to make high quality metareasoning decisions.

2.5   Summary

In this chapter we considered the first step in the metareasoning engineering process. At this point we should decide which metareasoning processes the meta-level will control and select the modes (what the meta-level will decide). There are many options, but we can consider the object level's computational limits or problems as potential targets for metareasoning. Naturally, some trial and error might be necessary before we can find a successful metareasoning approach.

The next step is to determine how to implement the metareasoning approach on the robot. Chapter 3 describes some options for locating the software that will implement the metareasoning process, but the details of the implementation depend upon the autonomy software architecture.

References

Arkin, R.C.: Behavior-Based Robotics. The MIT Press, Cambridge, Massachusetts (1998)

Bertsekas, D.P., Tsitsiklis, J.: Neuro-dynamic Programming. Athena Scientific, Belmont, Massachusetts (1996)

Boddy, M., Dean, T.L.: Deliberation scheduling for problem solving in time-constrained environments. Artif. Intell. **67**(2), 245–285 (1994)

Carrillo, E., Jaffar, M.K.M., Nayak, S., Patel, R., Yeotikar, S., Azarm, S., Herrmann, J.W., Otte, M., Xu, H.: Communication-aware multi-agent metareasoning for decentralized task allocation. IEEE Access **9**, 98712–98730 (2021)

Caron, P.: A simple introduction to meta-reinforcement learning. https://medium.com/instadeep/a-simple-introduction-to-meta-reinforcement-learning-6684f4bbd0de (2021). Accessed 9 Feb 2023

Cserna, B., Ruml, W., Frank, J.: Planning time to think: metareasoning for on-line planning with durative actions. In: Proceedings of the International Conference on Automated Planning and Scheduling, vol. 27, pp. 56–60 (2017)

Dasari, V.R., Geerhart, B.E., Alexander, D.M., Shires, D.R.: Distributed computation offloading framework for the tactical edge. In: IEEE INFOCOM 2019-IEEE Conference on Computer Communications Workshops (INFOCOM WKSHPS), pp. 1–6 (2019)

Dawson, M.K.: Metareasoning approaches to thermal management during image processing. Thesis, University of Maryland (2022)

Dawson, M.K., Herrmann, J.W.: Metareasoning approaches for thermal management during image processing. In: Proceedings of the ASME 2022 International Design Engineering Technical Conferences and Computers and Information in Engineering Conference, IDETC/CIE2022, St. Louis, Missouri, 14–17 Aug 2022

Etzioni, O.: Tractable decision-analytic control. In: Brachman, R.J., Levesque, H.J., Reiter, R. (eds.) Proceedings of the First International Conference on Principles of Knowledge Representation and Reasoning, pp. 114–125. Morgan-Kaufmann Publishers, San Mateo, California (1989)

Garey, M.R., Johnson, D.S.: Computers and Intractability: A Guide to the Theory of NP-Completeness. W.H. Freeman and Company, New York (1979)

Herrmann, J.W.: Engineering Decision Making and Risk Management. Wiley, Hoboken, New Jersey (2015)

Herrmann, J.W.: Data-driven metareasoning for collaborative autonomous systems. Technical Report, Institute for Systems Research, University of Maryland, College Park. http://hdl.handle.net/1903/25339 (2020). Accessed 10 Nov 2022

Jarin-Lipschitz, L., Liu, X., Tao, Y., Kumar, V.: Experiments in adaptive replanning for fast autonomous flight in forests. In: 2022 IEEE International Conference on Robotics and Automation (ICRA). Philadelphia, Pennsylvania, 23–27 May 2022

Karpas, E., Betzalel, O., Shimony, S.E., Tolpin, D., Felner, A.: Rational deployment of multiple heuristics in optimal state-space search. Artif. Intell. **256**, 181–210 (2018)

Langlois, S.T., Akoroda, O., Carrillo, E., Herrmann, J.W., Azarm, S., Xu, H., Otte, M.: Metareasoning structures, problems, and modes for multiagent systems: a survey. IEEE Access. **8**, 183080–183089 (2020)

Lee, J., Wang, P., Xu, R., Dasari, V., Weston, N., Li, Y., Bagchi, S., Chaterji, S.: Virtuoso: video-based intelligence for real-time tuning on SOCs. https://arxiv.org/abs/2112.13076 (2021). Accessed 15 Apr 2022

Lin, C.H., Kolobov, A., Kamar, E., Horvitz, E.: Metareasoning for planning under uncertainty. In: Twenty-fourth International Joint Conference on Artificial Intelligence (2015)

Molnar, S.L., Mueller, M., MacPherson, R., Rhoads, L., Herrmann, J.W.: Metareasoning to improve global and local path planning for a mobile ground robot. Technical Report, Institute for Systems Research, University of Maryland, College Park. http://hdl.handle.net/1903/29723 (2023)

Murphy, R.R.: Introduction to AI Robotics, 2nd edn. The MIT Press, Cambridge, Massachusetts (2019)

Navardi, M., Mohsenin, T.: MLAE2: Metareasoning for latency-aware energy-efficient autonomous nano-drones. In: 2023 IEEE International Symposium on Circuits and Systems (ISCAS). Monterey, California (2023)

Parashar, P., Goel, A.K., Sheneman, B., Christensen, H.I.: Towards life-long adaptive agents: using metareasoning for combining knowledge-based planning with situated learning. Knowl. Eng. Rev. **33**, 1–17 (2018)

Rabiee, S., Biswas, J.: IV-SLAM: introspective vision for simultaneous localization and mapping. In: Fourth Conference on Robot Learning (2020)

Russell, S., Wefald, E.: Do the Right Thing. The MIT Press, Cambridge, Massachusetts (1991a)

Russell, S., Wefald, E.: Principles of metareasoning. Artif. Intell. **49**(1–3), 361–395 (1991b)

Song, H., Peng, W.Z., Kim, J.H.: Partition-aware stability control for humanoid robot push recovery. In: Proceedings of the ASME 2022 International Design Engineering Technical Conferences and Computers and Information in Engineering Conference (IDETC-CIE 2022), St. Louis, Missouri, 14–17 Aug 2022

Sung, Y., Kaelbling, L.P., and Lozano-Pérez, T.: Learning when to quit: meta-reasoning for motion planning. In: 2021 IEEE/RSJ International Conference on Intelligent Robots and Systems (IROS), pp. 4692–4699 (2021)

Svegliato, J.: Metareasoning for planning and execution in autonomous systems. Dissertation, University of Massachusetts, Amherst (2022)

Svegliato, J., Wray, K.H., Witwicki, S.J., Biswas, J., Zilberstein, S.: Belief space metareasoning for exception recovery. In: 2019 IEEE/RSJ International Conference on Intelligent Robots and Systems (IROS), pp. 1224–1229 (2019)

Svegliato, J., Sharma, P., Zilberstein, S.: A model-free approach to meta-level control of anytime algorithms. In: 2020 IEEE International Conference on Robotics and Automation (ICRA), pp. 11436–11442 (2020)

Zilberstein, S.: Operational rationality through compilation of anytime algorithms. Dissertation, University of California, Berkeley (1993)

Zilberstein, S., Russell, S.: Approximate reasoning using anytime algorithms. In: Natarajan, S. (ed.) Imprecise and Approximate Computation, pp. 43–62. Springer, Boston, Massachusetts (1995)

Implementing Metareasoning

After deciding on the metareasoning problem and the metareasoning mode, the next step is implementing the metareasoning approach (cf. Fig. 3.1). The details of the implementation depend upon the autonomy software architecture, so Sect. 3.1 will review some of these architectures. Section 3.2 will discuss the options for locating the metareasoning policy in the architecture.

At this point, we're still unconcerned about the actual metareasoning policy. The implementation will provide the metareasoning functionality to which we'll later add the specific values of thresholds, rules, and other details of the metareasoning policy. Chapter 4 will discuss approaches for determining these details.

This chapter is also unconcerned with specifics of programming languages and operating systems. The approaches that we'll discuss here can be implemented in any modern programming language and operating system, so it doesn't matter if we use C++ or Python or something else.

3.1   Autonomy Software Architectures

The history of robotics includes various approaches for (and debates about) the proper architecture for the software that controls a robot's actions. One natural approach is to use a deliberative decision-making process: gather information from sensors and other sources, identify and evaluate the options, select the "best" one, and execute it. Deliberative architectures based on this process emphasized planning and decision making, but the decision latency (time lag) was sometimes excessive because it takes significant computational resources to solve complicated planning problems. Thus, robots that used these architectures sometimes moved and acted slowly. (There is a reason why the word "deliberate" means not only "thoughtful" but also "slow.")

J. W. Herrmann, *Metareasoning for Robots*, Synthesis Lectures on Computer Science, https://doi.org/10.1007/978-3-031-32237-2_3

Fig. 3.1 The metareasoning engineering process has steps that lead to one or more metareasoning policies and evidence about their performance. Although shown as a sequence of steps, it is often necessary to return to a previous step to repair issues that arise

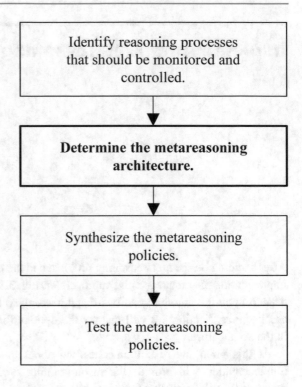

Reactive architectures that use simpler procedures to select actions quickly were the reasonable response to this, but those architectures don't integrate knowledge about the world because they assume (a) that the real-time data from the robot's sensors are adequate for the task and (b) that the limitations of the environment prevent valid deliberation (Arkin 1998). To overcome the limitations of deliberative architectures and reactive architectures, hybrid architectures were used to exploit the best of both architecture types.

We should note here the distinction between the "levels" of the canonical metareasoning framework (ground level, object level, and meta-level) and the "layers" of an autonomy software architecture. Whether the robot uses a reactive or deliberative or hybrid architecture, the layers in the autonomy software architecture are all elements in the object level; these layers do not correspond to the levels of the canonical metareasoning framework. For example, an assembly robot that uses a mission planner, a task planner, and specialized skill planners (cf. Parashar and Goel 2021) is using a hierarchical planning approach with multiple layers, but all of the planners are elements of the object level.

It can be useful to describe an autonomy software architecture using a set of simple verbs (Murphy 2019). These verbs indicate, at a high level of abstraction, how the architecture works.

We can describe a deliberative architecture as SENSE–PLAN–ACT. That is, the robot's sensors collect data about the environment (SENSE). Then the robot's software uses the sensor data (and other information that it has) to develop a plan for the robot (PLAN) and specific commands to the robot's motors and other actuators so that the robot can move somewhere or manipulate an object (ACT). Some deliberative architectures (which might be called hierarchical control or intelligent control) have a hierarchy of planners (or layers) that range from real-time actuator control to strategic global planning (Arkin 1998). The Shakey robot was an early example of this architecture (Kortenkamp et al. 2016).

We can describe a reactive architecture as SENSE–ACT. That is, the robot's sensors collect data about the environment (SENSE). Then the robot's software uses the sensor data (and other information that it has) and immediately sends specific commands to the robot's motors and other actuators so that the robot can move somewhere or manipulate an object (ACT). The subsumption architecture (Brooks 1986) is a well-known example that includes a set of layers; each layer focuses on one type of robot behavior, such as avoiding objects, exploring the environment, or identifying objects. Behaviors at a higher layer (such as obstacle avoidance) can override signals from other behaviors (such as exploration).

A hybrid architecture includes both reactive processes and deliberative processes (Brachman 2002). This can be described as "PLAN, then SENSE-ACT" (Murphy 2019). This architecture decouples long-term planning from real-time execution so that the planning process, which can take more time, doesn't degrade the robot's ability to react quickly. The 3T architecture uses an executive layer called the Sequencer as the interface between the reactive and deliberative layers; the Sequencer both determines the set of behaviors needed to accomplish the tasks that the planner selected and monitors the performance of these behaviors (Murphy 2019). A variety of three-tiered architectures have been developed (Kortenkamp et al. 2016).

The hybrid architecture presented by Kelly (2013) also has three layers: the reactive autonomy layer controls the robot's motion; the perceptive autonomy layer uses a local map and plans only a few seconds into the future; the deliberative autonomy layer uses a global map and does long-term mission planning.

The Soar architecture (Laird et al. 2012) is a different type of hybrid architecture: its reactive layer includes perception processes and controllers that can react to the data from the perception processes. Its deliberative layer includes a symbolic working memory and a set of rules and procedures that determine which actions the robot should perform.

An important part of the architecture is the communication approach that enables the autonomy software's modules or components to exchange data and instructions. There are two popular approaches: (1) client–server and (2) publish-subscribe (Kortenkamp et al. 2016). The client–server approach benefits from clearly defined interfaces, but the interfaces require significant overhead.

The publish-subscribe approach simplifies communication by including "topics" that serve as many-to-many communication channels. The autonomy software's modules or components can publish data or instructions to one or more topics and can subscribe to one or more topics to receive data or instructions asynchronously. The Robot Operating System (ROS) uses this approach (Cousins 2010; Open Robotics 2022).

Developing a metareasoning approach requires understanding the robot's autonomy software architecture, whatever it might be. After understanding that, we can consider the problem of where (in which part of the autonomy software) to implement the metareasoning approach.

3.2   Locating Metareasoning

As described in Chap. 1, we normally describe metareasoning as the function of a meta-level that is separate from the robot's object level. This is an accurate model of metareasoning, but it is not necessary to implement the meta-level as a separate computational process. It might be easier and better to implement the metareasoning policy in the same code that is running the object level reasoning processes. (Here, "better" means less programming effort and less overhead when running.) In other cases, that type of implementation might require much more effort, however.

First, we'll describe three different arrangements for locating metareasoning; these are shown graphically in Fig. 3.2.

The first case separates metareasoning from the autonomy software. The autonomy software (which performs the object level reasoning processes) and the metareasoning process run in parallel, as shown in Fig. 3.2a. They interact by sharing data or information of some kind, which depends on the nature of the metareasoning problem and the metareasoning mode (discussed in Chap. 2). The metareasoning process can be run on a separate set of hardware, or it can use the same set of computational resources that are performing the autonomy software. With this arrangement, the structure of the software

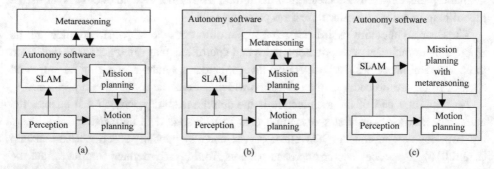

Fig. 3.2 Three arrangements for locating metareasoning with respect to the autonomy software; this figure shows a simplified autonomy stack and assumes that the meta-level is monitoring and controlling the mission planning module. **a** Metareasoning is a separate process. **b** Metareasoning is a module in the autonomy software. **c** Metareasoning is part of the mission planning module

most closely resembles the metareasoning framework because it has separate object level and meta-level processes.

The second case locates metareasoning as a module added to the autonomy software, as shown in Fig. 3.2b. The metareasoning process might be part of a layer in the autonomy software architecture, but that depends upon (1) the metareasoning problem and the metareasoning mode and (2) how the autonomy software architecture itself is implemented. This arrangement might allow the metareasoning process to access more information about the object level reasoning processes more quickly or directly. The meta-level is still defined by a specific component, but it is part of the autonomy software. This also holds if the robot is using multiple metareasoning processes, each of which is implemented as a module in the autonomy software.

The third case locates metareasoning inside the code of the autonomy software, as shown in Fig. 3.2c. The specific location depends on the nature of the metareasoning problem and the metareasoning mode. Some part of the autonomy software code is modified to include a metareasoning step (e.g., using a metareasoning policy to make a decision or set a parameter value) that is executed with the other code that performs the object level reasoning. In this arrangement, the meta-level is no longer separate; it is fully integrated into the autonomy software.

In the first case, the autonomy software includes only the object level reasoning processes. In the second and third cases, the autonomy software includes both the object level and the meta-level.

Wherever we decide to locate the meta-level, we should follow, as much as possible, the architecture style that the rest of the autonomy software is using. Employing the same style will make it easier for us to implement, test, and debug the meta-level, and it will reduce the likelihood of introducing a problem that might cause the autonomy software to malfunction.

If the meta-level is making an algorithm selection decision, then it might be reasonable to add a metareasoning routine to the object-level program that will make the object-level decision (such as which route to take to the destination). The object-level program has two critical steps: (1) call the metareasoning routine to select an algorithm and (2) run the selected algorithm. This is reasonable because the meta-level decision must be made only when the object level decision is needed. Dawson and Herrmann (2022) studied a metareasoning policy for image processing that was implemented this way.

Sung et al. (2021) used an anytime motion planning algorithm and included a metareasoning decision in each iteration of the algorithm. The robot used a metareasoning policy to determine whether it should continue the anytime motion planning algorithm or not. The metareasoning decision is a step in the iteration of the algorithm. A similar implementation was used by Jarin-Lipschitz et al. (2022), who described a motion planner with a metareasoning policy that adjusted a parameter that affected its solution quality and computational effort.

In the approach described by Goel et al. (2020), the metareasoning procedure is part of a four-phase planning process. The process assesses any problems in the planning algorithm and uses metareasoning to update the set of networks that the planning algorithm can use. The metareasoning procedure is implemented as a module in the robot's autonomy software.

If the meta-level is monitoring an anytime algorithm (denoted as A), however, it might be reasonable to implement the metareasoning routine as a separate process that periodically observes the current solution that A has found, measures the quality of that solution, predicts the future performance of A, and interrupts A if appropriate. Svegliato et al. (2018) described a metareasoning approach that was implemented this way for monitoring a tour improvement algorithm for solving the traveling salesman problem (TSP), a genetic algorithm for solving the job shop scheduling algorithm, and a simulated annealing algorithm for solving the quadratic assignment problem (QAP).

Parashar and Goel (2021) used a hybrid architecture for assembly robots and implemented metareasoning using two different structures. The architecture included a deliberative layer (for long-term planning), an executive layer, and a behavioral layer (for real-time reasoning). The deliberative layer of the architecture includes the mission planner and the task planner. For a product that needs to be assembled, the mission planner generates a partially ordered plan of part sequencing tasks. The task planner then uses hierarchical task network (HTN) planning to sequence the assembly tasks. The executive layer responds to task failures by invoking a repair heuristic, which implements metareasoning by adding to the object level reasoning process. The architecture's behavioral layer includes the specialized skill planners that convert the task into assembly primitives. The robot also has a separate metareasoning module that decides when to invoke a task repair algorithm that sends modifications to the task planner in the deliberative layer.

Greenwald (1996) added metareasoning to an avionics system and implemented it by adding another computer that ran the metareasoning procedure and communicated with the avionics system via the system bus. This arrangement corresponds to the separate metareasoning process in Fig. 3.2a.

To improve the performance of a local search motion planning algorithm, Cserna et al. (2017) implemented a metareasoning policy directly in the planning algorithm so that the meta-level had direct access to information about the planning options. The metareasoning policy determined when the planner should select an option that gave it more time for planning the next step; that is, the meta-level slowed the system (incurred a small delay) to allow better decision making, which allowed for a better plan that reached the goal in less time. This arrangement corresponds to the implementation in Fig. 3.2c.

The implementation of the metareasoning approach can significantly affect the computational burden of metareasoning (the metareasoning "overhead"). As described in this section, there are multiple implementation approaches. Thus, it is important to consider the options carefully and to select one that will allow the metareasoning approach to execute quickly. Doing this will keep the metareasoning overhead low and avoid degrading the robot's reasoning processes and its overall performance.

3.3   Case Study

Recent work at the University of Maryland implemented metareasoning in an autonomy stack that was installed on a Clearpath Jackal robot (shown in Fig. 3.3). We installed a RGB camera, a LIDAR, an IMU, and a GPS receiver on the robot. As part of the process of developing this metareasoning approach, we considered and tested multiple approaches for implementing metareasoning in the autonomy stack (Molnar et al. 2023). We'll first describe the approach that we implemented in the autonomy stack.

The robot's autonomy software was the Army Research Laboratory's ground autonomy stack (U.S. Army 2022), which includes subsystems for perception, simultaneous localization and mapping (SLAM), symbolic planning, and metric planning. The perception pipeline accepts inputs from the camera and LIDAR and includes modules for object detection, object tracking, object localization (with pose estimation), semantic segmentation, and similar processes that manipulate the sensor inputs. The SLAM subsystem includes modules for odometry, SLAM, rendering terrain, and sensor fusion. The symbolic planning subsystem includes modules for natural language understanding, mission planning, and planning behaviors. The metric planning subsystem includes modules for global planning, local planning, navigation, and control. Multiple global planners and multiple local planners are available. The software uses the Robot Operating System (ROS) (Open Robotics 2022), which enables the various modules to communicate via publish-subscribe message passing.

At the time that we performed this work, the ARL ground autonomy stack included four global path planners and three local path planners that were added by ARL researchers and their collaborators. (We did not create or modify these planners.) The global path planners can find a path from the robot's current location to the goal location within a relatively static map. To do this, they use different algorithms, including A*, generalized lazy search, and RRT* (Hedegaard et al. 2021; Likhachev 2010; Mandalika et al. 2019).

The local path planners determine the best way for the robot to move along the global path using information from its immediate environment, allowing the robot to navigate around dynamic obstacles and obstacles that the robot has only recently observed. The local plan operates independently to the waypoint goal. These planners use different algorithms, including nonlinear optimization, model predictive control, and A* (Howard et al. 2010; Howard and Kelly 2007; Wagener et al. 2019).

We conducted simulation experiments using a similar but larger robot that was running the same ground autonomy stack. Our preliminary results showed that the path planners

Fig. 3.3 A Clearpath Jackal

sometimes failed to generate a feasible plan, which led to a mission failure. The failures occurred for several reasons. If a path planner is being used in an environment that is not conducive to its functionality, then the algorithm will likely struggle to find feasible solutions that allow the robot to navigate over the terrain and around the obstacles. The second reason a path planner might fail is when an algorithm uses a predictive heuristic function that is unaware of future obstacles in the path. As the ground robot comes across a new obstacle, the planning algorithm might not have enough time to find a new plan before the ground robot collides with the new obstacle.

Figure 3.4 shows an example of a simulated robot that is stuck behind some trees in a forest. The robot is surrounded by many obstacles, and the path planning algorithm is unable to find a feasible path to the next waypoint. Because of the path planning failure, the robot cannot complete its mission. Motivated by such undesirable events, we investigated whether metareasoning can overcome path planning failures.

Because the ARL ground autonomy stack uses ROS to implement its reasoning processes, we decided to add the meta-level alongside these object level reasoning processes.

Fig. 3.4   A still image from a computer simulation in which a mobile ground robot is stuck behind some trees because its planning algorithms failed. (Image created by Sidney Molnar)

This corresponds to the second case that Sect. 3.2 listed: metareasoning is done by a module that is added to the autonomy software, as shown in Fig. 3.2b. This implementation enables the meta-level to monitor the data that the reasoning processes generate and to control the reasoning processes directly. In particular, we decided that, whenever a path planner failed, the meta-level should perform algorithm selection and switch from the current planner to a new one.

The "sequential" approach that we implemented adds a metareasoning node to the autonomy stack. The metareasoning policy is a predetermined sequence of global and local planner combinations. When the metareasoning node receives a stuck or error message produced by one of the planners, it selects the planner combination that is next in the sequence and then starts that global planner and local planner through a new launch file. The metareasoning node can do this because it is a ROS node that subscribes to the ROS topics to which the planners publish. Thus, it will receive any messages that indicate that a planner has failed. Moreover, it can issue the ROS commands that start the new nodes (for the new planners). By launching new planner nodes under the same name as the old planner nodes, the original nodes are automatically killed and replaced by the new nodes. Figure 3.5 demonstrates the logic used for the sequential approach.

Four global planners (SBPL, GLS, EASL, and RDGP) and two local planners (NLOPT and MPPI) were used in the sequential approach. We tested these combinations using simulation. For each run of the simulation, we recorded whether the robot failed to reach the

Fig. 3.5 A flowchart of the logic that the sequential metareasoning approach uses

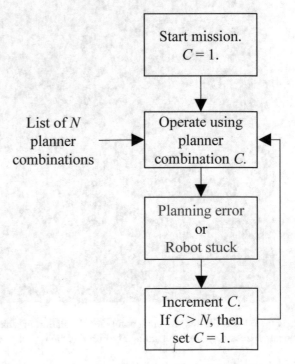

Table 3.1 The sequential TS (time-to-success) metareasoning policy

Priority	Global planner-local planner combination
1	EASL-NLOPT
2	EASL-MPPI
3	RDGP-MPPI
4	GLS-MPPI
5	GLS-NLOPT

Table 3.2 The sequential SR (success rate) metareasoning policy

Priority	Global planner-local planner combination
1	GLS-MPPI
2	EASL-MPPI
3	EASL-NLOPT
4	GLS-NLOPT
5	SBPL-MPPI

goal location (a mission failure) or (if it were successful) the time needed to reach the goal. The "success rate" is the number of successes divided by the number of runs. The "time to success" is the average time for successful runs. We selected the five combinations that had the best time to success and used those for one metareasoning policy, which will be known as the Sequential TS policy (Table 3.1). We also selected the five combinations that had the best success rate and used those for a second metareasoning policy, which will be known as the Sequential SR policy (Table 3.2). Within each policy, we sequenced the planner combinations from best performance to worst performance. Chapter 4 provides more information about the synthesis of this metareasoning policy.

We also tried implementing a different metareasoning approach. This implementation, which we called the "parallel" approach, also added metareasoning nodes to the autonomy stack. The key difference was that it used only one global planner but ran multiple local planners simultaneously. The meta-level had two nodes; the first, which ran the metareasoning policy, monitored the local planners, determined which ones were working, and selected the highest priority local planner that was working. It then told a "multiplexer" node which planner had been selected at that time, and the multiplexer node passed that planner's output to the other nodes that needed it. The outputs from the other local planners (the ones that were not selected) were ignored.

Section 4.2 discusses the process of creating the metareasoning policies, and Chap. 5 provides details about the testing that we did.

3.4   Summary

In this chapter we considered the second step in the metareasoning engineering process. In this step we should decide whether we will implement the metareasoning processes separately from the autonomy software or as part of the autonomy software. This chapter discussed these options and their advantages and disadvantages. We also examined a case study that illustrates how we implemented metareasoning as part of a ROS-based autonomy stack.

The next step is to synthesize the metareasoning policy that will control the object level reasoning process. Chapter 4 describes a systematic, data-driven approach for determining the metareasoning policy and presents some case studies to illustrate the process.

References

Arkin, R.C.: Behavior-Based Robotics. The MIT Press, Cambridge, Massachusetts (1998)

Brachman, R.J.: Systems that know what they're doing. IEEE Intell. Syst. **17**(6), 67–71 (2002)

Brooks, R.: A robust layered control system for a mobile robot. IEEE J. Robot. Autom. **2**(1), 14–23 (1986)

Cousins, S.: Welcome to ROS topics. IEEE Robot. Autom. Mag. **17**(1), 13–14 (2010)

Cserna, B., Ruml, W., Frank, J.: Planning time to think: metareasoning for on-line planning with durative actions. In: Proceedings of the International Conference on Automated Planning and Scheduling, vol. 27, pp. 56–60 (2017)

Dawson, M.K., Herrmann, J.W.: Metareasoning approaches for thermal management during image processing. In: Proceedings of the ASME 2022 International Design Engineering Technical Conferences and Computers and Information in Engineering Conference, IDETC/CIE2022, St. Louis, Missouri, 14–17 Aug 2022

Goel, A.K., Fitzgerald, T., Parashar, P.: Analogy and metareasoning: cognitive strategies for robot learning. In: Lawless, W.F., Mittu, R., Sofge, D.A. (eds.) Human-Machine Shared Contexts, pp. 23–44. Academic Press, London (2020)

Greenwald, L.: Analysis and design of on-line decision-making solutions for time-critical planning and scheduling under uncertainty. Dissertation, Brown University (1996)

Hedegaard, B., Fahnestock, E., Arkin, J., Menon, A., Howard, T.M.: Discrete optimization of adaptive state lattices for iterative motion planning on unmanned ground vehicles. In: 2021 IEEE/RSJ International Conference on Intelligent Robots and Systems (IROS), pp. 5764–5771 (2021)

Howard, T.M., Green, C.J., Kelly, A.: Receding horizon model-predictive control for mobile robot navigation of intricate paths. In: Howard, A., et al. (eds.) Field and Service Robotics, vol. 7, pp. 69–78 (2010)

Howard, T.M., Kelly, A.: Optimal rough terrain trajectory generation for wheeled mobile robots. Int. J. Robot. Res. **26**(2), 141–166 (2007)

Jarin-Lipschitz, L., Liu, X., Tao, Y., Kumar, V.: Experiments in adaptive replanning for fast autonomous flight in forests. In: 2022 IEEE International Conference on Robotics and Automation (ICRA). Philadelphia, Pennsylvania, 23–27 May 2022

Kelly, A.: Mobile Robotics. Cambridge University Press, New York (2013)

Kortenkamp, D., Simmons, R., Brugali, D.: Robotic systems architectures and programming. In: Siciliano, B., Khatib, O. (eds.) Springer Handbook of Robotics, pp. 283–306. Springer, Berlin (2016)

Laird, J.E., Kinkade, K.R., Mohan, S., Xu, J.Z.: Cognitive robotics using the Soar cognitive architecture. In: Workshops at the Twenty-Sixth AAAI Conference on Artificial Intelligence, Cognitive Robotics (2012)

Likhachev, M.: Search-based planning with motion primitives. https://mirror.umd.edu/roswiki/attachments/Events(2f)CoTeSys(2d)ROS(2d)School/robschooltutorial_oct10.pdf (2010). Accessed 10 Nov 2022

Mandalika, A., Choudhury, S., Salzman, O., Srinivasa, S.: Generalized lazy search for robot motion planning: interleaving search and edge evaluation via event-based toggles. In: Proceedings of the International Conference on Automated Planning and Scheduling, vol. 29, pp. 745–753 (2019)

Molnar, S.L., Mueller, M., MacPherson, R., Rhoads, L., Herrmann, J.W.: Metareasoning to improve global and local path planning for a mobile ground robot. Technical Report, Institute for Systems Research, University of Maryland, College Park. http://hdl.handle.net/1903/29723 (2023)

Murphy, R.R.: Introduction to AI Robotics, 2nd edn. The MIT Press, Cambridge, Massachusetts (2019)

Open Robotics: Documentation. https://wiki.ros.org/ (2022). Accessed 6 Aug 2022

Parashar, P., Goel, A.K.: Meta-reasoning in assembly robots. In: Systems Engineering and Artificial Intelligence, pp. 425–449. Springer, Cham (2021)

Sung, Y., Kaelbling, L.P., Lozano-Pérez, T.: Learning when to quit: meta-reasoning for motion planning. In: 2021 IEEE/RSJ International Conference on Intelligent Robots and Systems (IROS), pp. 4692–4699 (2021)

Svegliato, J., Wray, K.H., Zilberstein, S.: Meta-level control of anytime algorithms with online performance prediction. In: Proceedings of the Twenty-Seventh International Joint Conference on Artificial Intelligence (IJCAI-18) (2018)

U.S. Army: SARA CRA overview. https://www.arl.army.mil/business/collaborative-alliances/current-cras/sara-cra/sara-overview/ (2022). Accessed 11 Nov 2022

Wagener, N., Cheng, C.-A., Sacks, J., Boots, B.: An online learning approach to model predictive control. In: Robotics: Science and Systems (2019)

Synthesizing Metareasoning Policies

Controlling object level reasoning requires a metareasoning policy, which is a mapping from the state space to the set of reasoning options. The metareasoning policy can be a mathematical function, include a set of rules, or take some other form. Essentially, a metareasoning policy says "If this condition occurs, then change the object level reasoning in this way."

This chapter presents a systematic, data-driven approach for determining the metareasoning policy. In this approach, we first characterize the performance of the reasoning options under different conditions. We then determine the "best" reasoning option for each condition and use that information to synthesize the metareasoning policy. The synthesis is done "offline"; the robot then uses the metareasoning policy "online" (during its mission) to control the object level reasoning in real time with little overhead. Naturally, we might develop multiple metareasoning policies and then do additional testing to evaluate their performance. Chapter 5 will discuss the process of testing metareasoning policies.

As part of the metareasoning engineering process (Fig. 4.1), the synthesis activity is a type of detailed design activity. We've already selected the metareasoning problem and the metareasoning mode (as discussed in Chap. 2) and decided how to implement the metareasoning approach (cf. Chap. 3). We now need to add the finishing touches by generating the metareasoning policy that will be implemented. Although they are determined last, the details of the metareasoning policy are critically important, for they will control the robot's object level, which determines the robot's performance. Thus, it is important to have a data-driven approach that provides the rationale for the metareasoning policy.

Synthesis (discussed in this chapter) and testing (Chap. 5) are distinct activities, but they both involve collecting data from simulation experiments or field experiments. The key difference between these activities is that synthesis gathers data about the performance of specific reasoning options (such as motion planning algorithms) in order to

© The Author(s), under exclusive license to Springer Nature Switzerland AG 2023
J. W. Herrmann, *Metareasoning for Robots*, Synthesis Lectures on Computer Science,
https://doi.org/10.1007/978-3-031-32237-2_4

Fig. 4.1 The metareasoning engineering process has steps that lead to one or more metareasoning policies and evidence about their performance. Although shown as a sequence of steps, it is often necessary to return to a previous step to repair issues that arise

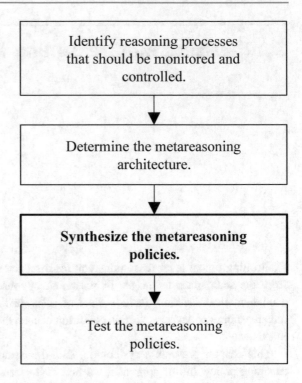

generate metareasoning policies, while testing evaluates the performance of the metareasoning policies. The simulation models and testbeds used for synthesis can be used again in testing.

To illustrate the synthesis process, this chapter presents some case studies that are based on published accounts and our own work. These indicate different ways to apply the general approach. These do not exhaust the possibilities, and there is plenty of room for creative approaches and policies.

Although there are many forms that a metareasoning policy can take, we can identify two common types: (1) policies that directly specify the preferred reasoning option (such as rules and classifiers) and (2) policies that evaluate a set of reasoning options (using performance profiles or regression functions) and select the "best" one. Of course, both types yield the same thing (a preferred reasoning option), but the different types influence the synthesis process, as we'll see.

Note that we're not asking the robot's meta-level to formulate and solve an optimization problem in real time. In general, this would require too much overhead and would interfere with the robot's ability to reason and react quickly when needed. Instead, we're specifying a metareasoning policy that the robot can use to control the object-level reasoning.

4.1   Synthesis Approach

This section presents a systematic but general approach for synthesizing and testing metareasoning policies offline. ("Offline" refers to computational work that is done before the robot begins operating and starts its mission.) This approach presumes that metareasoning will be implemented as a metareasoning policy that will select a reasoning option. This includes but is not limited to algorithm selection. The metareasoning policy uses the values of the state variables to select a reasoning option from a set of possible reasoning options.

The synthesis approach has three steps (shown in Fig. 4.2):

1. Definition: Define the state variables and performance measures; identify the reasoning options; define the domain of relevant instances.
2. Characterization: Conduct analysis, experiments, or simulations to collect data about the performance of the reasoning options across the domain of relevant instances and the set of possible states.

Fig. 4.2 The metareasoning synthesis approach has three steps that lead to one or more metareasoning policies

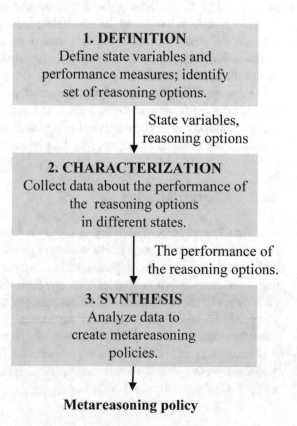

3. Synthesis: Use these results to determine which reasoning options are best in which situations and create one or more metareasoning policies.

The following subsections will discuss these steps briefly and give some examples from the literature, but these steps make more sense as an integrated process, so it will be useful to consider the examples of the case studies in Sect. 4.2, which illustrate the entire synthesis approach.

4.1.1 Definition

The first step in the synthesis approach resembles problem formulation. The state variables describe the information that is available. Let X be the space of values that the state variables can have. The performance measures identify the objective or objectives that we wish to achieve. The reasoning options are based on the metareasoning mode, which we discussed in Chap. 2. Let R be the set of reasoning options. The domain of relevant instances helps bound the scope of the metareasoning problem.

Possible state variables include the elapsed computation time, the quality of the current solution, the status of a reasoning algorithm (e.g., whether it has failed), the availability of communication, the status of computational resources, features of the environment, and the locations of other robots.

As discussed in Chap. 5, performance measures can describe the object level, ground level, or the entire system. These include solution quality (e.g., path length), prediction accuracy, utilization of computational resources (e.g., CPU, GPU, and memory), mission success, mission completion time, and the frequency of unsafe actions.

4.1.2 Characterization

An effective metareasoning policy is based on knowledge about the quality of the reasoning options as a function of the state, and characterization is the process of obtaining such knowledge. Analysis, experimentation, and simulation can be used for characterization. Here analysis is a general term for any logical or mathematical argument that yields insights about the performance of the reasoning options. Experimentation and simulation yield data about the performance of the reasoning options, which we must analyze to identify trends and develop inferences that can be used to create metareasoning policies. As discussed more in Chap. 5, the experimentation and simulation testing should occur in a controlled environment so that the impact of each reasoning option can be determined without interference from other factors.

4.1.3   Synthesis

The synthesis step creates the metareasoning policy, which can be considered as a mapping $p{:}X \to R$; that is, given the values of the state variables, the metareasoning policy p determines a reasoning option.

A wide range of approaches can be used in the synthesis step. In some cases, the performance of the reasoning options will yield patterns that clearly show which reasoning option is best in which states. In other cases, statistical analysis might yield insights into which state variables affect the performance of the reasoning options, and those results might suggest a certain rule as the metareasoning policy. In some cases, it might be necessary to use machine learning approaches to generate the performance estimation (regression) functions or classifiers that will form the metareasoning policy.

If we derive the metareasoning policy by analyzing the metareasoning decision, then we should understand that the resulting policy depends upon the assumptions that we made when formulating the decision problem. As Karpas et al. (2018) noted, reasonable assumptions that might be invalid in practice can be useful for formulating a helpful model and synthesizing an effective metareasoning policy; this phenomenon is related to George Box's statement that "all models are wrong, but some are useful."

If the characterization step does not yield results that lead directly to an "obvious" metareasoning policy, then we might consider a family of policies that share the same essential logic but have different thresholds. These can be tested to determine which threshold values lead to the best overall performance. (Chapter 5 discusses testing in more detail.)

If we suspect that the range of situations that the robot might encounter will change in the future, we might add a reinforcement learning approach to update the metareasoning policy over time (as discussed in Chap. 2).

Although one might use analysis or data for inspiration, generating effective metareasoning policies is essentially a creative activity, not a logical deduction. The case studies in Sect. 4.2 illustrate this principle.

4.2   Case Studies

Based on published accounts and our own work, this section presents some case studies of synthesizing metareasoning policies. Some are examples of metareasoning heuristics, and some are examples of learning a fixed metareasoning policy. Section 4.3 will discuss the similarities and differences of these cases.

4.2.1 Motion Planning and Path Optimization

Motion planning is an important reasoning process for an autonomous robot, and many approaches have been developed. Asymptotically optimal methods converge to an optimal solution as the number of samples increases infinitely (Karaman and Frazzoli 2011). In applications, however, sampling-based approaches that can quickly return solutions can be used. Because the best solution found is (likely) not optimal, path optimization techniques can be employed to improve the solutions.

1. Definition. Luna et al. (2013) considered a motion planning approach that uses path optimization techniques to improve the solutions found by sampling-based approaches for different motion planning problems. In this approach, a fixed time budget is allocated for motion planning, and the approach iterates until time expires; then the best solution found is returned. Each iteration calls the motion planner and a path optimization technique.

The metareasoning mode is algorithm selection, and the reasoning options are two path optimization techniques: shortcutting (Geraerts and Overmars 2007) and hybridization (Raveh et al. 2011). The metareasoning policy decides which path optimization technique to use in each iteration. Because the computation time was fixed, the key performance measure was the length of the path.

2. Characterization. Their discussion of the two path optimization techniques provides some insights into how Luna et al. characterized the reasoning options and developed their metareasoning policy. Shortcutting is a "micro-optimization" procedure that refines a single path by removing suboptimal motions, but it cannot find paths in a new homotopy class, which limits the amount of improvement possible. (If two solutions are in the same homotopy class, then it is possible to transform one into the other through a continuous function; see Fig. 4.3 for an example.) Hybridization, however, is a "macro-optimization" procedure that can discover solutions in a new homotopy class, which might yield a bigger improvement. Here, Luna et al. did not use experimentation or simulation to characterize the reasoning options. Instead, they understood the techniques' strengths and weaknesses from previous work and created a metareasoning heuristic.

3. Synthesis. From this characterization, Luna et al. synthesized a metareasoning policy that simply alternates between the two reasoning options. The only state variable is the parity of the number of iterations performed. The metareasoning policy selects shortcutting if the state is even and hybridization if the state is odd. (That is, it switches back and forth between these two options.) We can express this metareasoning policy as the following mapping:

p:{even, odd} $\rightarrow$ {shortcutting, hybridization}.

p("even") = "shortcutting." p("odd") = "hybridization."

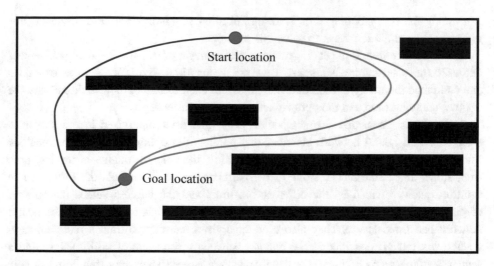

Fig. 4.3 In this path planning example, three feasible paths from the start location to the goal location avoid the obstacles (the black rectangles). The orange and green paths (on the right) are in the same homotopy class; the red path (on the left) is in a different homotopy class

4.2.2   Stopping Anytime Motion Planning

As we discussed in Chap. 2, stopping an anytime algorithm is a common metareasoning problem. Sung et al. (2021) synthesized multiple metareasoning approaches for stopping an anytime motion planning algorithm. They studied three different motion planning algorithms: RRT*, PRM*, and Lazy PRM*. Here we'll focus on their use of neural networks as metareasoning polices that specify a reasoning option as a function of the algorithm's current performance.

1. Definition. The anytime motion planning algorithm first found and then improved a feasible path; after a feasible path was found, the reasoning options at each time step are "continue" and "stop." The objective is to maximize utility, which is a function of solution quality and computation time. The solution quality is the length of the path (in a two-dimensional space). Initially, the utility increases as the motion planning algorithm finds much better (shorter) paths, but utility then decreases as the improvements in solution quality become smaller and smaller but more time elapses. Solution quality was normalized relative to the length of the initial feasible path and the estimated optimal path length for the current problem. Let q be the normalized solution quality; this ranges from 0 to 1. Let t be the normalized time; this also ranges from 0 to 1. Let w be the weight that expresses the decision-maker's preferences about the tradeoff between solution quality and time. Then, let U be the utility of a path that has quality q found at time t:

$$U(q, t; w) = wq + (1 - w)t$$

Finding the best stopping time is an optimization problem; Chapter 2 discussed this type of policy and other ways to solve this problem.

2. Characterization. Sung et al. generated a random set of two-dimensional workspaces and used those as training examples. They ran the anytime algorithm on these examples and computed the utility at each time step; thus, for each example, they could identify the optimal stopping time and the optimal actions at each time step.

3. Synthesis. From this characterization, they then used supervised learning to train a feed-forward neural network classifier that predicts the optimal action ("continue" or "stop") given the time, the current solution quality, the solution quality at the last time step, slope, and flatness. The slope of the performance profile equals the difference in solution quality divided by the difference in time, and the flatness equals the number of previous time steps for which the solution quality value has not changed. The neural network had three layers. They also used supervised learning to train a recurrent neural network (RNN) that models the relation between a sequence of utility values and a sequence of reasoning options. (For details, see their paper.) Their approach is an example of learning a fixed metareasoning policy, which Chap. 2 discussed.

4.2.3 Selecting Collaboration Algorithms

The benefits of multi-robot systems (MRS) come from their ability to solve tasks more efficiently through collaboration. Some of the key challenges in MRS, such as the coordination of robots' behavior and distributed task allocation, require communication between robots. In realistic environments, however, communication can be unreliable and outside the control of the MRS, which makes task coordination more challenging.

Carrillo et al. (2021) and Nayak et al. (2020) investigated the problem of distributed task allocation in environments with variable communication availability. For example, the communication availability might change from low (where most transmitted messages are not received) to high (where most transmitted messages are received). This project focused on a MRS in which the robots were performing missions motivated by search-and-rescue and similar settings and were collaborating to complete the mission as quickly as possible. Their approach relied upon offline characterization and synthesis to generate a metareasoning policy that the robots used in real time.

1. Definition. In this work, the state space X was the availability of communication between the robots (measured by the likelihood that a transmitted message will be received). The set R of reasoning options included the following collaboration algorithms: the Consensus Based Auction Algorithm (CBAA), the Asynchronous Consensus Based Bundle Algorithm (ACBBA), the Decentralized Hungarian Based Algorithm (DHBA), the Hybrid Information and Plan Consensus (HIPC) algorithm, and the Performance Impact (PI) algorithm.

An instance i described the locations of the robots and the locations that the robots needed to visit (these locations were called "tasks"). Using reasoning option r in R (one of the collaboration algorithms) generated the solution, a set of task allocations that determined which location(s) each robot should visit. This task allocation problem was solved repeatedly during a mission until all locations were visited.

The relevant performance measures were the number of transmitted messages (lower is better) and the distance traveled (lower is better).

2. Characterization. The collaboration algorithms were tested in multiple scenarios under different states (communication conditions) to determine the quality of the solutions (how quickly the robots completed the mission) and the amount of communication required (the number of messages transmitted). The evaluation of the algorithms was done using computer simulation that simulated not only the movement of the robots in the environment but also the communication between them (Nayak et al. 2020). This allowed the evaluation in states with different levels of communication quality; the simulation model used the Rayleigh fading model, which includes two factors that attenuate the signal strength: fading and path loss, which depends upon the distance from the transmitter to a receiver (see Carrillo et al. for details).

3. Synthesis. The research team then analyzed the results and directly designed a metareasoning policy that was expressed as a small set of rules. They used reactive synthesis, a type of formal method, to generate a distributed multi-agent metareasoning policy from specifications written in Linear Temporal Logic (LTL). This switching protocol determined which collaboration algorithm should be used based on the perceived communication availability. Each robot estimated the communication availability based on signals from other robots and used the metareasoning policy to select a collaboration algorithm (the reasoning option); it then used this collaboration algorithm to determine its task sequence assignment. Different policies were created for different scenarios (search & rescue, wildfire monitoring, and ship protection). For example, for the ship protection scenario, the metareasoning policy selected the performance impact algorithm when communication quality is high and the decentralized Hungarian-based algorithm otherwise.

4.2.4 Simultaneous Localization and Mapping

1. Definition. Simultaneous Localization and Mapping (SLAM) is an important reasoning process in a robot's object level. Piazza et al. (2022) conducted simulation experiments to test the performance of three SLAM packages that can be used in the ROS (Robot Operating System) framework: GMapping, SLAM Toolbox, and Hector SLAM. The simulations covered four indoor locations with rooms and layouts of different sizes, and the simulated robot was a Turtlebot 3 Waffle (Open Source Robotics n.d.). (For additional details, please see their paper.) From these results, they fitted generalized univariate and

multivariate linear regression models to estimate each package's performance when used with different sensors and in different environments. They considered five performance measures: the average CPU utilization, the maximum memory utilization, the normalized translation relative error, the normalized rotation relative error, and the absolute trajectory error. The key features of the sensors are the maximum range of the laser scan, the width of the field of view, and the noise (error) in the odometry. The key feature of the environment around the robot is its self-similarity, which is expressed as the translation geometric similarity.

2. Characterization. Piazza et al. simulated 2460 runs of the robot through the four environments (layouts) to obtain the training data. They first generated univariate models, which yielded insights into general trends of SLAM performance (for example, the translation error of the GMapping algorithm degrades quickly when the laser scan range becomes too small, but it degrades more gradually as the field of view narrows). They also generated multivariate models, one for each performance measure, which yields insights into how particular combinations of features affect the relative performance of the SLAM packages. For instance, when the field of view was 180°, the odometry was perfect (no noise) and the translation geometric similarity was very high, the GMapping algorithm and the SLAMToolbox had similar translation error when the laser scan range was less than 8 m, but the GMapping algorithm performed better (less translation error) when the laser scan range was greater than 8 m.

3. Synthesis. Although the particular prediction models that Piazza et al. developed are valid only for the SLAM packages, hardware, and environments that they studied, this data-driven approach generates functions that can be used by a metareasoning policy that performs algorithm selection. For instance, if we consider the fact that the SLAMToolbox uses fewer computational resources (memory and CPU) than GMapping, we might propose a metareasoning rule (heuristic) that selects SLAMToolbox except for those cases when the GMapping yields significantly lower error.

4.2.5 Global and Local Path Planning

This case study is based on some of our recent work on metareasoning for a mobile ground robot; this was also discussed in the case study in Chap. 3. For additional details not discussed here, see Molnar et al. (2023).

1. Definition. The Army Research Laboratory's ground autonomy stack (U.S. Army 2022) includes subsystems for perception, simultaneous localization and mapping (SLAM), symbolic planning, and metric planning. The metric planning subsystem includes modules for global planning and local planning.

Preliminary simulation experiments with the ground autonomy stack showed that the path planners sometimes failed to generate a feasible plan, which led to a mission failure. This led to the decision to add a meta-level that can recover from a failure by switching the global and local path planning algorithms. The state variables are the status of the planners (failed or not) and which planners are currently active. The reasoning options are the different combinations of global and local path planners. In this approach, the metareasoning mode includes both when to switch path planners and algorithm selection.

2. Characterization. Molnar et al. tested four global planners (SBPL, GLS, EASL, and RDGP) and two local planners (NLOPT and MPPI). They tested multiple combinations in three simulation scenarios (see Figs. 4.4, 4.5, and 4.6). In each run of the simulation, if the robot failed to reach the goal location, this was recorded as a mission failure. If it were successful, the time needed to reach the goal was recorded. The "success rate" is the number of successes divided by the number of runs. The "time to success" is the average time for successful runs.

3. Synthesis. Based on the results, Molnar et al. selected the five combinations that had the best time to success and used those for one metareasoning policy, which was called the Sequential TS policy (cf. Table 4.1). They also selected the five combinations that had the best success rate and used those for a second metareasoning policy, which

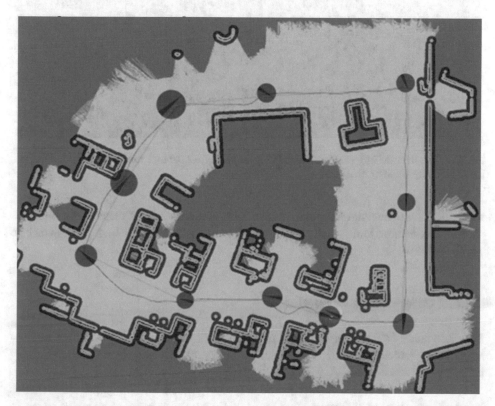

Fig. 4.4 The least challenging scenario for the metareasoning test. (*Image credit* Robbie MacPherson, Lawrence Rhoads, Matt Mueller)

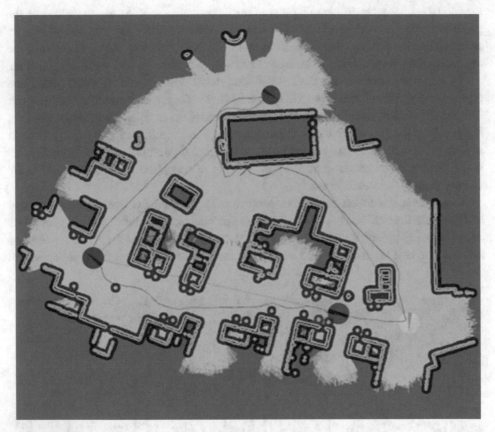

Fig. 4.5 The intermediate scenario for the metareasoning test. (*Image credit* Robbie MacPherson, Lawrence Rhoads, Matt Mueller)

was called the Sequential SR policy (Table 4.2). The sequence of planner combinations within each metareasoning policy was set by ordering them from best performance to worst performance.

4.2.6 Defending a Perimeter

This case study is based on some of our recent work on metareasoning for the perimeter defense problem. Prannoy Namala and Arjun Vaidya developed these metareasoning policies.

1. Definition. Perimeter defense games, a type of pursuit evasion game, are scenarios in which one or more defenders must try to minimize the number of attackers trying to enter a target region. The defenders can move but must stay on a given perimeter (a closed

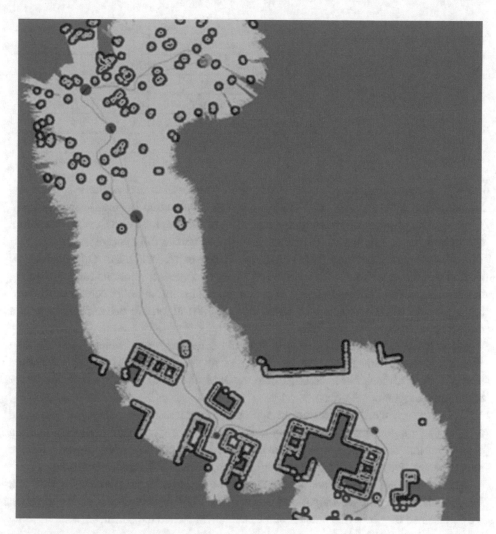

Fig. 4.6 The most challenging scenario for the metareasoning test. (*Image credit* Robbie MacPherson, Lawrence Rhoads, Matt Mueller)

Table 4.1 The sequential TS (time-to-success) metareasoning policy	Priority	Global planner-local planner combination	Average time to success (s)
	1	EASL-NLOPT	179.46
	2	EASL-MPPI	195.04
	3	RDGP-MPPI	212.95
	4	GLS-MPPI	239.58
	5	GLS-NLOPT	240.63

Table 4.2 The sequential SR (success rate) metareasoning policy

Priority	Global planner-local planner combination	Average success rate (%)
1	GLS-MPPI	70
2	EASL-MPPI	68
3	EASL-NLOPT	63
4	GLS-NLOPT	53
5	SBPL-MPPI	38

curve in a two dimensional space). There are many civilian and military applications for perimeter defense games, and this is a potential application for multi-robot systems. Typically, the problem of assigning the defenders to intercept the attackers is solved in a centralized manner so that the defenders' actions can be carefully coordinated.

The Maximum Matching (MM) algorithm (Chen et al. 2017) and the Local Game Region (LGR) algorithm (Shishika et al. 2020) are two general solutions for coordinating the defenders. The LGR algorithm will find an optimal solution under certain conditions, but it requires more computational effort than the MM algorithm does. Thus, there is a tradeoff between the two algorithms.

We decided to use metareasoning to help the defenders select which algorithm they should use during an attack. The key idea was to use the LGR algorithm only when it might yield a much better solution than the MM algorithm (which would justify the additional computational effort).

2. Characterization. To develop an appropriate metareasoning policy, we generated a dataset of random instances, where the defenders were distributed randomly along a circular perimeter, and the attackers were distributed randomly throughout the area outside the perimeter. We also varied the number of defenders and the number of attackers.

After generating these instances, each one of these instances are simulated twice: with the defenders using the MM algorithm and with the LGR algorithm. For each simulation we determined the score (the number of attackers that reached the perimeter without being intercepted by a defender).

Using the score value, the instances are divided into three classes: A, B, and C. Class A includes the instances where both algorithms yielded the same score. Class B includes the instances where the score for the LGR algorithm is less than the score for the MM algorithm. Class C includes the instances where the score for the MM algorithm is less than the score for the LGR algorithm. (The defenders wish to minimize the score.)

3. Synthesis. We used the results from the two datasets to train various machine learning models that we then used as metareasoning policies. The models were classifiers that predicted the class of an instance based on the number of defenders, the number of attackers, and features related to the defenders' and attackers' locations. The metareasoning policy then selected an algorithm based on the predicted class: LGR for class B, and

MM for classes A and C (because, if we expect both algorithms to have the same score, we will choose the one that requires less computational effort).

4.3 Discussion

The case studies that we covered in Sect. 4.2 had many similarities. The flowchart in Fig. 4.7 shows a common pattern. The engineers started with a set of reasoning options and then used simulation models to evaluate the performance of the options. They then analyzed the results and generated some rules that the metareasoning policy used to control the object level reasoning.

In some of the cases, the engineers used machine learning techniques to develop a classifier that became part of the metareasoning policy. We did this for the perimeter defense problem, as shown in Fig. 4.8. In other cases, the engineers reviewed the results and then directly specified one or more rules (heuristics) for the metareasoning policy.

The motion planning and path optimization case study (Luna et al. 2013) used a different approach, however. They used their understanding of the algorithm's strengths and weaknesses to specify a rule instead of conducting a simulation study to evaluate the reasoning options.

For the global and local path planning case study, we later devised a different approach that would require identifying a set of "situations" in which the planners failed, testing the various global planner-local planner combinations in each situation, and determining the

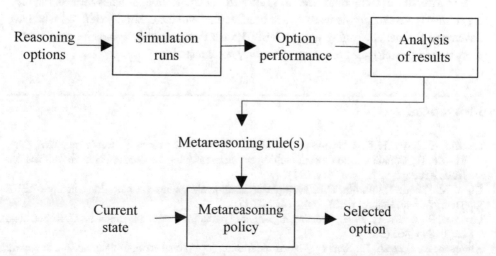

Fig. 4.7 A schematic of the approach for synthesizing a metareasoning policy by evaluating the reasoning options. When the robot is operating, the metareasoning policy uses the metareasoning rules to select a reasoning option

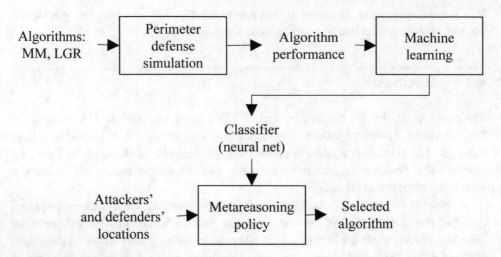

Fig. 4.8 A schematic of our approach for synthesizing a metareasoning policy for the perimeter defense problem. The metareasoning policy determines which algorithm the defenders should use to coordinate their behavior. MM = Maximum matching. LGR = Local game region

best combination for each situation. Given this information, the corresponding metareasoning policy would, when the planners fail, identify the situation that most resembles the robot's current predicament and start the best global planner-local planner combination for that situation.

The diversity of techniques used to synthesize metareasoning policies reflects the variety of metareasoning problems that have been studied and the creativity of those who have developed the metareasoning approaches. The synthesis approach and the case studies discussed in this chapter are meant to inspire, not constrain.

References

Carrillo, E., Jaffar, M.K.M., Nayak, S., Patel, R., Yeotikar, S., Azarm, S., Herrmann, J.W., Otte, M., Xu, H.: Communication-aware multi-agent metareasoning for decentralized task allocation. IEEE Access **9**, 98712–98730 (2021)

Chen, M., Zhou, Z., Tomlin, C.J.: Multiplayer reach-avoid games via pairwise outcomes. IEEE Trans. Autom. Control **62**(3), 1451–1457 (2017)

Geraerts, R., Overmars, M.H.: Creating high-quality paths for motion planning. Int. J. Robot. Res. **26**(8), 845–863 (2007)

Karaman, S., Frazzoli, E.: Sampling-based algorithms for optimal motion planning. Int. J. Robot. Res. **30**(7), 846–894 (2011)

Karpas, E., Betzalel, O., Shimony, S.E., Tolpin, D., Felner, A.: Rational deployment of multiple heuristics in optimal state-space search. Artif. Intell. **256**, 181–210 (2018)

Luna, R., Şucan, I.A., Moll, M., Kavraki, L.E.: Anytime solution optimization for sampling-based motion planning. In: Proceedings of the IEEE International Conference on Robotics and Automation (ICRA), pp. 5068–5074, Karlsruhe, Germany, 6–10 May 2013

Molnar, S.L., Mueller, M., MacPherson, R., Rhoads, L., Herrmann, J.W.: Metareasoning to improve global and local path planning for a mobile ground robot. Technical Report, Institute for Systems Research, University of Maryland, College Park. http://hdl.handle.net/1903/29723 (2023)

Nayak, S., Yeotikar, S., Carrillo, E., Rudnick-Cohen, E., Jaffar, M.K.M., Patel, R., Azarm, S., Herrmann, J.W., Xu, H., Otte, M.W.: Experimental comparison of decentralized task allocation algorithms under imperfect communication. IEEE Robot. Autom. Lett. 5(2), 572–579 (2020)

Open Source Robotics: Turtlebot 3. https://www.turtlebot.com/turtlebot3/. Accessed 10 Nov 2022

Piazza, E., Lima, P.U., Matteucci, M.: Performance models in robotics with a use case on SLAM. IEEE Robot. Autom. Lett. 7(2), 4646–4653 (2022)

Raveh, B., Enosh, A., Halperin, D.: A little more, a lot better: improving path quality by a path-merging algorithm. IEEE Trans. Rob. 27(2), 365–371 (2011)

Shishika, D., Paulos, J., Kumar, V.: Cooperative team strategies for multi-player perimeter-defense games. IEEE Robot. Autom. Lett. 5(2), 2738–2745 (2020)

Sung, Y., Kaelbling, L.P., Lozano-Pérez, T.: Learning when to quit: meta-reasoning for motion planning. In: 2021 IEEE/RSJ International Conference on Intelligent Robots and Systems (IROS), pp. 4692–4699 (2021)

U.S. Army: SARA CRA overview. https://www.arl.army.mil/business/collaborative-alliances/current-cras/sara-cra/sara-overview/ (2022). Accessed 10 Nov 2022

Testing Metareasoning Policies

After considering the metareasoning design options (Chap. 2), selecting a metareasoning approach, implementing it (Chap. 3), and synthesizing one or more metareasoning policies (Chap. 4), we need to begin testing and evaluation. If we have only one candidate metareasoning policy, we must conduct tests to evaluate the robot's performance with and without metareasoning (cf. Fig. 5.1). Systematic testing can provide evidence that the metareasoning policy is reducing the computational resources required for reasoning, improving the robot's performance, reducing the risk of unsafe actions, or creating other benefits. Although we hope that testing will yield such evidence, we also know that a poor metareasoning policy might degrade performance in one or more ways, and testing can reveal that as well.

If we have multiple candidate metareasoning policies, then testing can determine which ones are superior and identify the tradeoffs that exist (if any).

In any case, testing might lead us to reconsider the metareasoning policy (or policies) and possibly the overall metareasoning approach. We'll then need to return to one of the prior steps, rework what we did before, and test again to determine if our changes have led to any improvements. This iterative approach is a typical development process, and it should be familiar to most engineers who work on robots. More generally, testing (both direct and vicarious) generates engineering knowledge and supports the selection of one design or approach over the other possibilities (Vincenti 1990).

This chapter discusses testing for evaluating a metareasoning approach and provides guidelines that can help engineers design test plans.

J. W. Herrmann, *Metareasoning for Robots*, Synthesis Lectures on Computer Science, https://doi.org/10.1007/978-3-031-32237-2_5

Fig. 5.1 The metareasoning engineering process has steps that lead to one or more metareasoning policies and evidence about their performance. Although shown as a sequence of steps, it is often necessary to return to a previous step to repair issues that arise

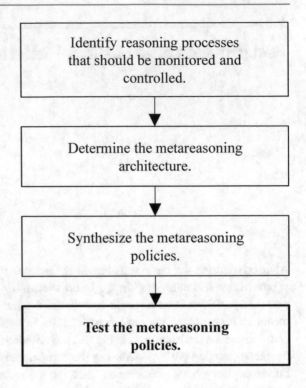

5.1   Performance Metrics

By implementing a metareasoning policy in the robot's meta-level, we hope to improve its object level reasoning and its ground level performance. Thus, when considering the performance of metareasoning, we should consider performance metrics that describe the impacts on all three levels.

Meta-level metrics. At the meta-level, the relevant performance metrics include those that capture what the metareasoning policy is doing and how many computational resources the metareasoning policy requires. For instance, if the metareasoning policy is performing algorithm selection, then we might count how many times each algorithm is selected.

For a problem that involved multi-robot collaboration, Herrmann (2020) used simulation to evaluate four collaboration algorithms and 42 metareasoning policies that selected one of the four algorithms based on the values of fourteen state variables. Figure 5.2 shows, for each policy, how frequently each policy chose each algorithm. Policy 31, the metareasoning policy that had the best team performance, frequently used one good, fast algorithm (Closest Target) and occasionally one better but more expensive one (CBAA).

If the metareasoning policy is adjusting a reasoning algorithm's parameter value, then we might track this value over time and note when and for how long the parameter

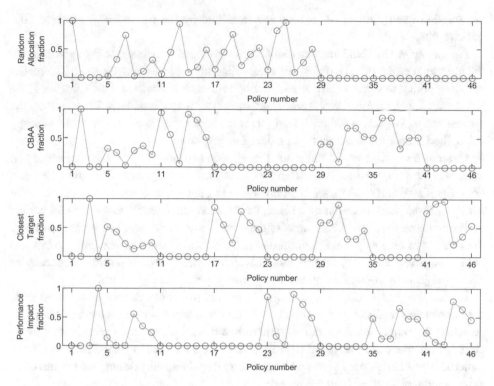

Fig. 5.2 The relative use of each candidate algorithm in each policy. 1 = Random allocation, 2 = CBAA, 3 = Closest target, 4 = Performance impact. See Herrmann (2020) for definition of each metareasoning policy (5–46)

has different values. Unfortunately, determining the actual computational burden of the metareasoning policy might be infeasible; it depends upon how the metareasoning policy is implemented (as discussed in Chap. 3) and the monitors that are available on the robot's computer and operating system.

Object level metrics. We can also track what the object level is doing, the computational resources that are required, and the quality of the solutions or decisions that the object level reasoning processes are generating. For instance, we might track how often different reasoning processes execute (this should correlate directly with the metareasoning decisions) and the computational resources (such as CPU utilization and memory) used by those reasoning processes. The details of monitoring and recording these metrics will depend upon the robot's computer and operating system. The quality of the solutions or decisions depends upon the type of reasoning process. For instance, if the object level reasoning process is a route planning algorithm, then the solution quality might be the distance of the path that it finds, the time required to traverse that path, or how frequently the algorithm fails to find a feasible solution. If it is an object detection algorithm, then the solution quality might be its accuracy measured by the true positive rate and false positive rate.

For example, Houeland and Aamodt (2018) developed a metareasoning approach that did algorithm selection. The object level reasoning problem was classification. The metareasoning policy selected a classification method for each instance in a data set. Thus, the key object level metric was classification accuracy. When tested on 21 data sets, the normalized classification accuracy using their metareasoning approach was (on average) 2.84% better than the accuracy of a random forest approach.

Ground level metrics. At the ground level, we can observe the robot's behavior and track what it does, whether it completes its mission, how long that takes, and whether it does anything unsafe. Often, we'll be interested in the time to complete the mission. For instance, if the robot is searching an area, the time to complete the search is an important performance metric. The effectiveness of the robot is also important. For a search mission, the number of objects that the robot found is a key performance metric. For a robot that is picking and placing objects, the rate at which it places objects and the frequency of placing objects in the right positions are important performance metrics.

In their study of distributed task allocation in multi-robot search missions with variable communication availability, Carrillo et al. (2021) developed a metareasoning policy that switched the robots' collaboration algorithm based on the current communication availability. They measured the total distance that the robots traveled to complete their mission (where lower is better) and used that when comparing the metareasoning policy and other collaboration algorithms (without metareasoning).

System level metrics. At the system level, we might be interested in both object level metrics such as computation time and ground level metrics such as path length. In that case, we can define a utility function that models our preferences about the trade-off between multiple measures. As discussed in Sect. 2.3, one approach for metareasoning is to optimize a utility function that is based on solution quality and computation time.

Sung et al. (2021) used this type of utility function in the metareasoning approaches that they developed for improving two-dimensional motion planning. The solution quality q was a measure of the path length relative to the worst path length (where $q = 0$) and the optimal path length (where $q = 1$). The utility function $U(q, t)$ was a linear combination of the solution quality q and the normalized computation time t. Their study is also notable because they tested five metareasoning policies (two of which were simple rule-based policies) and evaluated them against an "oracle" policy that could make optimal metareasoning decision.

Metareasoning should improve the performance of the robot, so it is valuable to describe how metareasoning changes the performance metrics. That is, we should evaluate the performance metrics both when the robot is operating without metareasoning and when it is using metareasoning. Expressing the performance impact of metareasoning as a relative change is often appropriate.

5.2 Test Planning

After identifying the performance metrics that we want to evaluate, we turn to planning the actual tests (experiments). This might repeat (or reuse) the analysis, simulation, or experimentation that was done to characterize the performance of the reasoning options, which was discussed in Chap. 4. At this point, however, we're interested in evaluating the robot's performance when using metareasoning (instead of evaluating individual reasoning options).

Murphy (2019) discussed different types of experiments, including controlled experimentation, exercises, concept experiments, and participant-observer studies. Both controlled experimentation and exercises are most relevant to our goals here.

Controlled experimentation. Controlled experimentation is used to test a hypothesis. For instance, does using the metareasoning policy reduce the utilization of the robot's CPU? Generally, we want to evaluate the performance quantitatively so that we can determine precisely the improvement (or degradation) due to the metareasoning policy. The testing should occur in a controlled environment so that the impact of the metareasoning policy can be determined without interference from other factors. Thus, we can perform a repeatable experiment that provides credible evidence about the robot's performance with and without the metareasoning policy (or with different metareasoning policies).

Typically, it is desirable to perform multiple experiments with different setups (for instance, different sets of obstacles that the robot must avoid) to determine how much the environment affects the relative performance of the metareasoning policy. Results that show that a metareasoning policy improves the robot's performance in some situations but not others can lead to ideas for modifying the metareasoning policy.

Simulation-based testing can be used to reduce the time and effort of controlled experiments. We can run experiments in simulated environments that are much larger and complicated than our research lab while still maintaining complete control, and these are repeatable as well. In general, there is a range of modeling approaches in addition to computer simulation, including analytical solutions to mathematical models, experiments with smaller scale physical models, and experiments with the actual system (Law 2007). For many robotics applications, however, simulation is less expensive, less disruptive, and safer. Fortunately, as computer hardware and software have become more powerful, we can build and run realistic simulation models and obtain valuable insights about a robot's behavior.

To test a safety metareasoning approach for a planetary rover, Svegliato et al. (2022) used a simulation to compare the performance of different versions of the rover, including one with no safety process. Clearly, real-world experiments were infeasible for testing this system. Even if a rover and a surrogate testbed (something that resembles the surface of another planet) were available, real-world testing that might damage the rover would be undesirable.

Carrillo et al. (2021) used simulation experiments to test the performance of a metareasoning policy that was developed for a multi-robot system (MRS). A MRS requires communication between robots to coordinate the robots' behavior and allocate tasks to the robots. In realistic environments, however, communication can be unreliable and outside the control of the MRS, which makes task coordination more challenging. The metareasoning approach monitored the communication availability and used algorithm selection to determine (based on the current communication availability) which collaboration algorithm the robots should use to allocate tasks among themselves. They developed metareasoning policies for three scenarios: search and rescue (with stationary targets), fire monitoring (with spreading targets), and ship protection (with moving targets). (The synthesis of the metareasoning policy was discussed in Chap. 4)

After developing the metareasoning policies, Carrillo et al. tested each one in a simulation framework that was constructed using ROS (Open Robotics 2022). The framework included modules for the simulated robots and a module for the environment (cf. Nayak et al. 2020). The environment module tracked the robots and the targets. It also changed the communication availability following a specific schedule. The robots communicated over the ROS network through a communication interface written in C++. This interface dropped messages randomly according to the specified communication model, which was the Rayleigh fading model. For each scenario, they randomly generated fifty instances of the problem with five to ten robots. The instances for the search and rescue scenarios and ship protection scenarios had up to forty targets. The instances for the fire monitoring scenario had up to ten initial fires. Simulation runs were done with the corresponding metareasoning policy and with the component algorithms. The communication availability changed during the simulation run, and two conditions were used: (1) low availability to high availability, and (2) high availability to low availability. Note that the simulation framework was designed to model imperfect communication, which was an important part of the problem. By using simulation, Carrillo et al. could control all aspects of the scenario, the dynamic communication availability, and the MRS mission.

Rabiee and Biswas (2020) used both simulation and real-world experiments to test their introspective visual SLAM (IV-SLAM) approach that can overcome the problems caused by reflections, glare, and shadows. In AirSim, a photo-realistic simulation environment, they simulated a car driving over 60 km in different conditions that challenged the SLAM process. They also implemented IV-SLAM on a Clearpath Jackal robot and drove it over 7 km in a variety of settings, including those that cause problems for visual SLAM. Note that using simulation allowed them to test IV-SLAM in bad weather conditions that they did not consider in their real-world experiments. Most researchers do not want to operate their robots in such conditions, which would possibly damage the robot. They conducted the simulation experiments both with and without the introspection function that enables metareasoning. They measured the mean reprojection error of the image features, tracking accuracy, and tracking failures.

For testing their adaptive motion planning algorithm, Jarin-Lipschitz et al. (2022) used a photorealistic simulation and flew an autonomous unmanned aerial vehicle (UAV) in a New Jersey state forest. In the simulation experiments, they could control the density of the trees and generate virtual forests in which the density varied, which was the motivation for their metareasoning approach. (Flying through a forest with an unchanging density would not require an adaptive motion planning approach.) They conducted simulation runs with and without metareasoning, which showed that the metareasoning approach could generate a successful plan in situations where the traditional planner could not.

Exercises. In an exercise, we test the robot in a more realistic environment following a scripted set of activities (Murphy 2019). Using a staged world and a script enables the test to be repeated, which was also important in the controlled experiments, but an exercise is more realistic, so it will likely be more expensive and time-consuming to do. If it is more difficult to collect the precise data that we can get in the lab, the results might be limited to a smaller set of performance metrics. Still, testing in a more realistic setting can help stakeholders see the potential benefits of using metareasoning more clearly, and the challenges (and failures) should provide insights that can be used for modifying the metareasoning policy.

Some organizations have constructed environments specifically for testing robots, and these are valuable locations for exercises, as they are designed to provide real-world challenges. For example, the Guardian Centers site in Perry, Georgia, has multi-story buildings, semi-collapsed structures that can be configured to replicate different degrees of collapse, tunnels, and other scenarios. These can be used for evaluating emergency response robots on standard tests while observers watch the exercise safely (Guardian Centers 2022).

In both controlled experimentation and exercises, it is important to conduct tests both with metareasoning and without metareasoning. Our goal should be to evaluate the impact of using metareasoning: How does metareasoning affect the robot's performance? How does the robot's performance change when we add metareasoning?

Component testing. When the metareasoning approach itself includes novel components, especially those that make predictions, we should do component-level testing to verify that their performance is adequate. If not, then we need to modify the approach and replace the bad components.

As part of their testing of a metareasoning approach for two-dimensional (2D) motion planning, Sung et al. (2021) used simulation studies for component-level testing. For instance, they tested the convolutional neural network (CNN) that was used to predict the shortest path length. For evaluating the path length prediction, they randomly generated 1000 2D workspaces of different types, used 800 for training the CNN, and used 200 for testing the CNN. The performance metric was the percent similarity of the predicted path length to the optimal path length.

5.3   Visualizing Metareasoning

While developing and implementing metareasoning, we'll find it useful to "see" what the robot is thinking. This can help us answer questions such as the following: How often does the metareasoning policy choose different reasoning options? Is the metareasoning policy choosing the "best" reasoning option? How is the performance of the object level changing?

Log files are a valuable resource for answering these questions. We can program the meta-level to record essential data about the state of the object level and the metareasoning choices every time it uses the metareasoning policy. The object level can also record the robot's location. By including timestamps in these log files, we can then synchronize them to link the metareasoning decisions to the robot's actions, verify that the metareasoning policy is working correctly, and investigate any malfunctions or unexpected behavior. If the robot's reasoning processes use ROS (Open Robotics 2022), then we can use rosbags to record the messages that the ROS nodes share via the ROS topics. We can then use Python or C++ libraries to read the data from these bags.

Another option is to monitor the robot's reasoning processes in real time. This resembles the telemetry that NASA uses to monitor a spacecraft during its mission. Using a direct radio channel or a wireless network, the robot can send information from its sensors such as its camera, LIDAR, and GPS to the command post computer. Software on that computer can display this information. For instance, it can show the video feed and a map with the robot's current position and pose (orientation), a path indicating where it has been, and current LIDAR returns. If the robot is using ROS, then we can use the rviz tool to visualize this information. Figure 5.3 is an example of the rviz output.

A useful example of visualizing metareasoning was presented by Rabiee and Biswas (2020), who developed the introspective visual SLAM (IV-SLAM) approach. They captured images from their real-world experiments and highlighted the reliable and unreliable features that were extracted by their approach. Their images show that IV-SLAM successfully handled shadows, reflections, glare, and other challenges.

Likewise, Jarin-Lipschitz et al. (2022) tracked the planning time as their UAV flew through the forest. When the forest became more cluttered, the planning time increased, and the metareasoning policy modified the dispersion, which decreasing the time required for planning. They generated a graph that shows the planning time, the dispersion parameter value, and the forest density; this image helps us visualize the metareasoning policy at work.

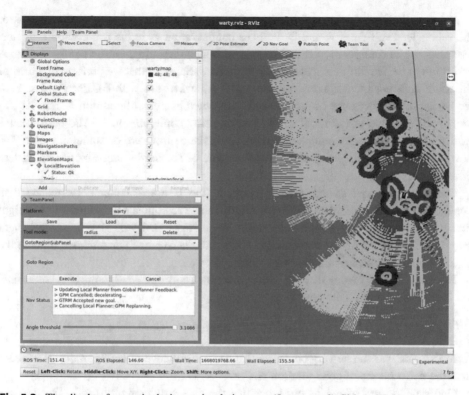

Fig. 5.3 The display from rviz during a simulation run. (*Image credit* Sidney Molnar)

5.4 Data Analysis

After completing the tests, we have data about the robot's performance. We then need to analyze the data to determine whether metareasoning improved performance (or determine which metareasoning policy performed best).

Although this section cannot thoroughly review all of the relevant statistical techniques, we can cover some useful methods and provide an illustrative example. For a comprehensive online guide to statistical methods, see the *e-Handbook of Statistical Methods* (NIST/SEMATECH 2012). In that guide, the techniques covered here can be found in Sect. 7.3 "Comparisons based on data from two processes." In addition, the simulation textbook by Law (2007) discusses techniques for analyzing the output of simulation experiments to compare and rank alternative system configurations. Although Law's work focuses on discrete-event systems (such as production lines), the data analysis approaches can be applied in many settings. Cohen (1995) covered a variety of useful hypothesis testing and estimation approaches and computational statistical methods such as Monte Carlo sampling, bootstrap sampling, and randomization tests. His book also covers more general topics related to testing AI algorithms.

Suppose that we're interested in comparing the performance of two configurations for the robot; let's call them A and B. Perhaps configuration A is the robot without metareasoning, while configuration B is the robot with metareasoning. Or the configurations A and B might be two different metareasoning policies. Suppose that we performed n runs of the robot under each configuration (n runs with A, and n runs with B). For $i = 1, \ldots, n$, let Y_{Ai} be the performance of the robot in run i when using configuration A, and let Y_{Bi} be the performance of the robot in run i when using configuration B. The performance metric might the mission completion time, the distance traveled, or something else. For now, we'll presume that we prefer a smaller value of the performance metric. (e.g., less time is better.)

Our key question is whether using configuration B led to better performance than configuration A, or vice versa, or their performance was equivalent. Naturally, we might first look at the average performance across the n runs. That will give us some insight into the performance of the two configurations, but the average performance ignores the distribution of performance across the n runs.

Now, one special case is that the worst performance of one configuration is better than the best performance of the other:

$$\max\{Y_{Bi}\} < \min\{Y_{Ai}\}$$

In this case, we can be very confident that the performance of configuration B is better than the performance of configuration A.

One standard test of means presumes that the data are "paired"; that is, the first run with configuration A is (except for the change in configuration) identical to the first run with configuration B, and the second runs are identical to each other, and so forth, through all n runs. Then, for $i = 1, \ldots, n$, we calculate the difference for each run:

$$d_i = Y_{Ai} - Y_{Bi}$$

The null hypothesis is that expected value of the difference equals 0; that is, changing the configuration does not change the robot's performance. Let α be the significance level, where α is chosen to be small (e.g., 0.01, 0.05, or 0.10).

We then find the average $\overline{d}$ and sample standard deviation s_d of these differences and use these to calculate the test statistic t:

$$t = \frac{\overline{d}}{s_d/\sqrt{n}}$$

We reject the null hypothesis if $|t| \geq t_{1-\frac{\alpha}{2}, n-1}$. If the performance with one configuration is much better than the performance with the other configuration, then the differences $d_1, \ldots, d_n$ will be large (either very positive or very negative), so the test statistic t will be large, and we'll reject the null hypothesis. On the other hand, if the two configurations yield similar performance, then the differences will be small (and likely both positive and negative), so the test statistic t will be small, and we cannot reject the null hypothesis.

A more general test is the Mann–Whitney U Test (NIST/SEMATECH 2012), which makes no assumptions about the distributions of the data. This test ranks the data instead of considering the differences of pairs.

We first rank the $2n$ values (all of the Y_{Ai} and Y_{Bi}). If two or more observations are tied, then each one is given the average of their ranks. For example, if three values are tied for the first three positions (ranks 1, 2, and 3), all three are given the rank 2. Let T_A be the sum of the ranks of the Y_{Ai}. Let T_B be the sum of the ranks of the Y_{Bi}. We then calculate the U statistics:

$$U_A = \frac{n}{2}(3n + 1) - T_A$$

$$U_B = \frac{n}{2}(3n + 1) - T_B$$

A larger value of the U statistic implies better performance (because the sum of the ranks is lower).

The null hypothesis is that the two sets of data (the Y_{Ai} and Y_{Bi}) have the same central tendency; that is, changing the configuration does not change the robot's performance. Let α be the significance level, where α is chosen to be small (e.g., 0.01, 0.05, or 0.10).

Now, we calculate the test statistic z:

$$z = \frac{\min\{U_A, U_B\} - n^2/2}{\sqrt{n^2(2n + 1)/12}}$$

We reject the null hypothesis if $|z| \geq z_{1-\alpha/2}$. If the performance with one configuration is much better than the performance with the other configuration, then the sums T_A and T_B (and U_A and U_B) will be very different, and the test statistic z will be large, and we'll reject the null hypothesis. Otherwise, the statistics will be very close, and the test statistic z will be close to zero, and we won't reject the null hypothesis.

As an example, let's consider some data that was collected in the study reported by Herrmann (2020). Configuration A will be a task allocation algorithm that is computationally intensive. Configuration B will be a metareasoning policy that sometimes chooses that algorithm and otherwise chooses a much simpler task allocation algorithm. Each configuration was evaluated with same $n = 100$ simulation runs. That is, the first simulation run for configuration A and the first simulation run for configuration B had the same starting positions for the robots and their adversaries. The performance metric is the number of successful attacks by the adversaries, which we want to minimize. Tables 5.1 and 5.2 list the results for these simulation runs. Because the results range from zero to five successes for both configurations, it is not clear that either one is better than the other.

For configuration A, the average performance is 1.52 successes; for configuration B, the average performance is 1.05 successes. It seems that using configuration B led to

Table 5.1 Performance (number of successful attacks) of configuration A in 100 simulation runs

Runs 1–20	Runs 21–40	Runs 41–60	Runs 61–80	Runs 81–100
1	1	0	1	1
4	2	3	2	2
1	0	1	1	2
1	2	2	2	0
1	1	5	4	1
0	0	1	3	1
0	1	2	1	2
2	3	3	2	3
2	0	0	0	4
1	3	0	1	1
0	1	1	0	3
1	3	3	1	1
0	1	5	0	1
3	1	1	0	3
2	0	0	2	3
3	3	0	1	1
0	1	2	0	1
2	1	4	2	1
1	2	0	4	1
1	1	3	1	3

better performance. Can we show that this is a statistically significance improvement? Or is it insignificant? Let the significance level $\alpha = 0.10$.

The average difference $\overline{d} = 0.47$ successes. The sample standard deviation $\sigma_d = 0.870$ successes. The test statistic $t = 5.40$, and the critical value $t_{0.9599} \approx 1.66$, so we reject the null hypothesis that configurations A and B have the same performance.

If we now use the Mann–Whitney U Test, we sort the data and determine the ranks. (Note that all 78 runs with no successes have the average rank of 39.5.) The sum $T_A = 11426.5$; the sum $T_B = 8673.5$. Therefore, $U_A = 3623.5$; $U_B = 6376.5$. The test statistic $z = -3.36$, and the critical value $z_{0.95} = 1.64$. Because $|z| \geq z_{0.95}$, we reject the null hypothesis that configurations A and B have the same performance.

The two statistical tests that we used provide evidence that using that metareasoning policy (Configuration B) improved the performance of the multi-robot system and that this improvement is statistically significant. Moreover, in this case, because that metareasoning policy often used a simpler task allocation rule, it also reduced the computational effort of task allocation.

Table 5.2 Performance (number of successful attacks) of configuration B in 100 simulation runs

Runs 1–20	Runs 21–40	Runs 41–60	Runs 61–80	Runs 81–100
0	0	0	0	0
4	0	0	0	0
1	0	0	0	2
0	0	2	2	0
0	0	5	4	0
0	0	1	3	1
0	0	2	1	1
0	0	3	2	0
2	0	0	0	4
1	3	0	1	0
0	0	0	0	3
0	3	0	0	0
0	1	5	0	0
3	1	0	0	3
2	0	0	3	3
3	3	0	0	0
0	0	4	0	1
1	1	4	2	3
0	0	0	4	0
0	1	3	0	3

5.5 Assurance Cases

An assurance case is a structured argument that a claim is true. Although assurance cases are often used in safety engineering to investigate claims about the safety of a new system, they can be used more generally. The top-level claim can be an assertion about some property of the robot (Bloomfield et al. 2021); for example, the top-level claim might be that a Mars rover can navigate to desired locations in the desired amount of time, conduct various tests, and transmit the results of these tests to scientists here on Earth (Mesmer 2022). Various approaches and tools are available for creating and expressing assurance cases, including the AdvoCATE tool (Denney and Pai 2018), the Claims, Argument, Evidence (CAE) approach (Bloomfield et al. 2021), and Goal Structuring Notation (GSN) (Assurance Case Working Group 2018; Hawkins et al. 2021).

The key concept is that the assurance case (often represented as a diagram) logically connects evidence (from tests, existing data, or other sources) to the claim that we wish to make. In a very simple case, the claim is quantitative, and the evidence directly shows that the claim is true. For example, if we claim that the robot's path planner can find a feasible path within one second, then evidence about its computation time can directly support that claim.

In other cases, however, the logical connection is indirect, and multiple types of evidence will be needed. For example, we might argue that the Mars rover will successfully execute tasks on Mars (a top-level claim) by showing that it can successfully execute those tasks on Earth (Mesmer 2022). We need evidence that this is a correct argument, and perhaps the only available evidence is the belief of a subject matter expert who has experience with Mars rovers. Furthermore, we need to support the claim that the rover can successfully execute those tasks on Earth. This involves more precise claims about rover speed, wheel strength, wheel slippage in sandy conditions, and other requirements. Supporting these claims will require evidence from different types of tests, simulation experiments, and engineering analyses.

More generally, an assurance case enables better decision making by explicitly showing the evidence and beliefs that support a claim. It does not logically and objectively prove the claim in the same way that a geometer can prove the Pythagorean theorem. It relies instead upon arguments and beliefs that we accept because they meet the standards of our community, which is another way of generating valid knowledge (Polanyi 1962).

We're interested in the performance of a metareasoning policy. Our top-level claim is the following: (T) the robot performs better when it uses metareasoning than it does without metareasoning. This is important but somewhat imprecise. For instance, what do we mean by "performance"? What do we mean by "metareasoning"? The next step is to make the top-level claim more precise in two ways: (A) the robot is more likely to reach a goal location when it uses a sequential metareasoning policy than when it uses a single path planning algorithm; (B) the robot can reach the goal location more quickly when it uses a sequential metareasoning policy than when it uses a single path planning algorithm. (We can make these even more precise by referring to specific environments or distances.)

We also claim that demonstrating claims A and B allows us to justify the top-level claim T (this is a side claim about the reasoning process and is, therefore, a type of metareasoning!). This might be based on a document (such as a list of requirements) that states that the likelihood of success and the time to complete the mission as the key metrics of performance. If we have such a document, it can be the evidence that justifies this claim. Figure 5.4 is a diagram of the assurance case that connects claims A and B to the top-level claim T. (Fig. 5.4 follows the Claims, Argument, Evidence (CAE) approach. One "reads" this type of figure from the evidence at the bottom towards the claim at the top.)

Claims A and B are more precise than T, and we can directly support (or rebut) them by collecting data about the robot's performance with the sequential metareasoning policy and with the individual path planning algorithms.

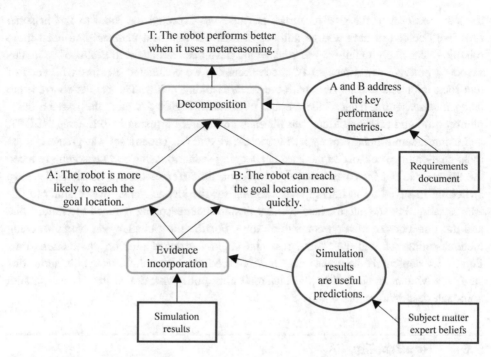

Fig. 5.4 An example of an assurance case diagram with claims (blue ovals), arguments (brown ovals), evidence (red rectangles), and reasoning blocks (green rectangles). (Claims have been abridged to save space)

If we can get data from experiments with actual robots, then we are likely done with our assurance case. If, however, we need to rely upon simulation experiments, then we need a side claim that the results of simulation experiments are sufficient (we don't need experiments with actual robots). The evidence for this might be a statement from a subject matter expert, or we might have data from previous work that show simulation results that were very close to the results from experiments with the actual robot in similar situations.

Case study. Our research group developed a metareasoning policy for an autonomous ground robot and planned to use a simulation study (a type of controlled experimentation) to evaluate it. The robot's meta-level controlled the reasoning process by employing multiple local planners that searched for routes around obstacles. The metareasoning policy determined when to switch from one local planner to another.

We developed three simulation scenarios through which the simulated robot needed to move. We deliberately used scenarios with different degrees of difficulty, based on the number and type of obstacles (these are shown in Figs. 5.5, 5.6, and 5.7). The most challenging scenario included navigating through a village and across hills with trees. We measured two performance metrics: the likelihood of failure (when the robot was unable to reach the goal location) and mission completion time (the time needed to move from the start location to the goal location). Because we expected the robot to fail in some runs, we decided to have a large number of runs. (If the local planners had been more reliable—less likely to fail—fewer runs would have been sufficient to evaluate the metareasoning policy's performance.) For each scenario, we conducted 30 runs with each of two local planners and 30 runs with the metareasoning policy. The results showed that using metareasoning reduced the number of failures. Figure 5.8 plots the mission completion time and mission failure rate for each configuration (using MPPI, using NLOPT, and using metareasoning) for each scenario (A, B, and C). The impact was greatest in the most challenging scenario. In the least challenging scenario, using metareasoning reduced the failure rate but increased mission completion time (for successful missions) because switching involved a short delay while waiting for the just-activated local planner to finish planning. For the intermediate scenario, using metareasoning reduced the failure rate and did not increase mission completion time. For the most challenging scenario, using metareasoning reduced the failure rate and slightly reduced mission completion time. Figure 5.4 depicts the assurance case that is based on the results. For details about this case, see Molnar et al. (2023); the metareasoning policy mentioned here is the parallel approach described in that report.

5.6   Replicability

We want to be confident that our test results are correct so that we can move forward with the development of our robot. Conducting multiple tests and getting similar results every time provides evidence that the test procedure is repeatable and that the metareasoning approach is truly successful. The ability to confirm the results of an experiment (or test) by repeating the experiment is known as *replicability* (National Academies of Sciences, Engineering, and Medicine 2019).

Replicability is closely related to "reproducibility," and the two terms are not used consistently across all disciplines . The National Academies report defines these terms as follows: reproducibility describes "obtaining consistent results using the same input data; computational steps, methods, and code; and conditions of analysis"; replicability is "obtaining consistent results across studies aimed at answering the same scientific question, each of which has obtained its own data." Because we're mostly concerned with using new experiments to confirm a result (not repeating the analysis of given data), we'll use the term "replicability" while noting that many discussions of reproducibility provide insights about replicability as well.

Within an organization, replicability is important because it allows one's colleagues to repeat a test in the future if needed to support the development of new versions or new approaches. In addition, future users, customers, or regulatory agencies might want to see the tests before purchasing or approving the system.

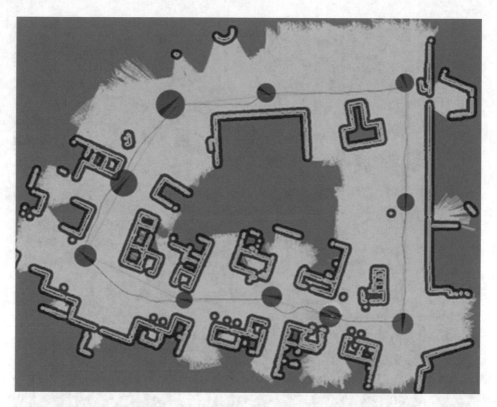

Fig. 5.5 The least challenging scenario for the metareasoning test. (*Image credit* Robbie MacPherson, Lawrence Rhoads, Matt Mueller)

More generally, in scientific research, reproducibility and replicability are valuable because it enables other researchers to confirm new results and to compare new approaches to existing ones more easily. For instance, suppose I develop a new path planning algorithm (NPPA) and show that NPPA can find better solutions more quickly than existing algorithms. To ensure replicability, I should publish the details of NPPA, including the source code, and information about the problem instances that I used for testing NPPA and the existing algorithms. If someone else subsequently develops a new version of the algorithm (say, NPPA*), then he can compare NPPA* and NPPA by using the same tests that I performed to evaluate NPPA in my work.

In general, two categories of problems make replicability and reproducibility difficult: human factors and technical issues (Dalle 2012). The human factors include insufficient detail in publications, ignorance of parameters and settings in the applications used, and mistakes that are made during the research. The technical issues including errors in the software, unavailability of software, obsolete software, and the use of floating point numbers to store and perform operations on integers.

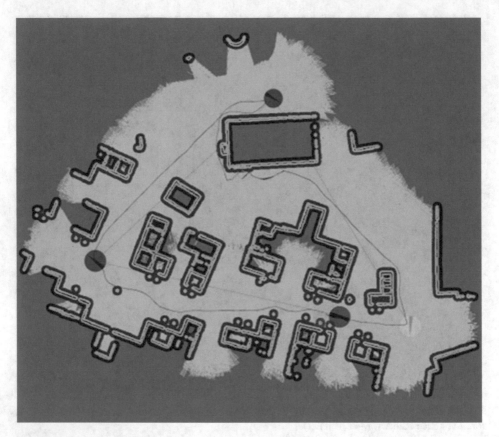

Fig. 5.6 The intermediate scenario for the metareasoning test. (*Image credit* Robbie MacPherson, Lawrence Rhoads, Matt Mueller)

Researchers should conduct their studies in ways that enable replicability and reproducibility in all areas of science, including computer science and simulation. The Association for Computing Machinery (ACM) published a list of actions that support reproducibility (Rous 2022), including the following:

- Clarify basic definitions, evaluation criteria, and branding;
- Enable sharing of artifacts (software, data, and results); and
- Preserve artifacts for future re-use.

Raghupathi et al. (2022) developed a reproducibility framework that groups 25 variables into three factors: Method, Data, and Experiment. This is meant to help researchers write papers that ensure reproducibility. Addressing all 25 variables should maximize the reproducibility of the research. Associated with each variable is a question, and these

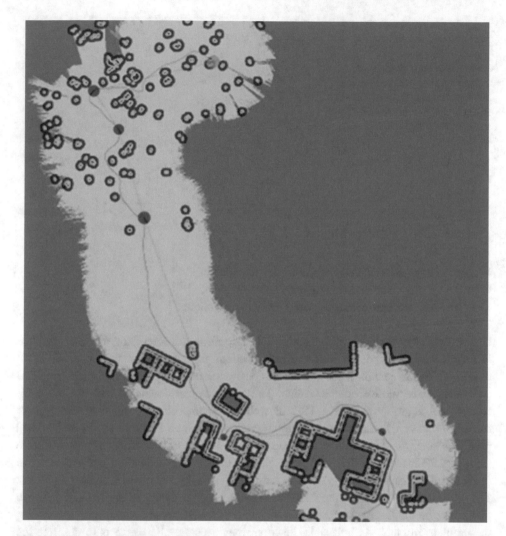

Fig. 5.7 The most challenging scenario for the metareasoning test. (*Image credit* Robbie MacPherson, Lawrence Rhoads, Matt Mueller)

form a checklist of the information that should be documented to ensure reproducibility. The following are some of the questions that are most relevant to testing metareasoning approaches:

- Is there an explicit mention of the algorithm(s) the research used?
- Are the variable settings shared, such as hyperparameters?
- Is there an explicit mention of scope and limitation of the research?
- Is the test data set shared?

Fig. 5.8 Scatter plot of the test
results. Each line corresponds
to a different planning
approach (a local planner or
metareasoning), and each point
corresponds to a scenario (least
challenging, intermediate, or
most challenging)

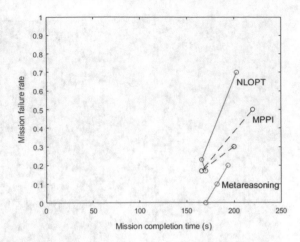

- Are the evaluation metrics of the model shared?
- Is the system code available as open source?
- Is the software used in the research explicitly mentioned?

One especially important but challenging part of replicability is re-using the same problem
instances (e.g., the specific start and end locations of a robot and the locations of all
obstacles). In some fields, researchers have constructed a standard set of problem instances
and those made those available online. Thus, anyone can download and use those for
testing new algorithms. On the TSPLIB website, an important example of this type of
resource, one can find instances of the Traveling Salesman Problem (TSP) and related
problems such as the Hamiltonian cycle problem and vehicle routing problems. (There
are 139 TSP instances.) The website (Reinelt n.d.) has documentation that explains the
format of the files that include the data and a FAQ to help users with common problems.
The documentation includes the formulas that one needs to calculate the distance between
two points on the Earth's surface (each specified by a latitude and a longitude). If sharing
the problem instances is infeasible, however, then a useful alternative is to describe the
details of the procedure used in to generate the test instances.

Replicability and reproducibility depend upon documentation and transparency. The
engineers who are planning the tests should document all aspects of the tests, whether
they are simulations with virtual robots or experiments with actual robots. They should
save the results of tests and the analysis that is done on those results. These details should
be consolidated into a single document or collection and archived. They should be shared
with others who might wish to review the tests and the results.

The desirability of replication studies depends upon various factors (National
Academies of Sciences, Engineering, and Medicine 2019). Conceptually, a replication
study is desirable if the expected value of the information that it provides is worth the
cost of running the study. If the previous result seems implausible, if there are doubts

about the correctness of the previous work, if the new results might settle an important controversy, or if the new results might have a large societal impact, then the expected value of the replication study is greater. On the other hand, if the replication study will require extensive money, time, or other resources, then it might not be worth the cost. Finally, performing a replication study has an opportunity cost because the resources that we use for that study cannot be used for other purposes that might generate more valuable results.

5.7   Summary

The results of testing can be uncertain: the results might show that our metareasoning approach has improved the robot's performance in some important ways, but they might show that the metareasoning approach is causing the robot's object level to perform poorly, which might lead us to redesign the metareasoning approach or to proceed without metareasoning.

After we define the relevant metrics, test planning determines the specifics of the test plans, including controlled experimentation and exercises. During and after running a test, visualizing metareasoning can help us understand what is happening. After collecting and analyzing the test results, we can use assurance cases to document the logical arguments that support our conclusions. Finally, we can document and share the testing approach and results to ensure replicability.

Ultimately, if we're successful, we have designed and implemented a useful metareasoning approach and have conducted testing to show that using it improves the robot's performance. The examples that we've discussed in these chapters have shown us how others have done this. We hope that your efforts to use metareasoning will be successful as well.

References

Assurance Case Working Group: Goal structuring notation community standard. Technical Report SCSC-141BA v2.0, Safety Critical Systems Club (2018)

Bloomfield, R., Fletcher, G., Khlaaf, H., Hinde, L., Ryan, P.: Safety case templates for autonomous systems. http://arxiv.org/abs/2102.02625 (2021). Accessed 13 May 2022

Carrillo, E., Jaffar, M.K.M., Nayak, S., Patel, R., Yeotikar, S., Azarm, S., Herrmann, J.W., Otte, M., Xu, H.: Communication-aware multi-agent metareasoning for decentralized task allocation. IEEE Access **9**, 98712–98730 (2021)

Cohen, P.R.: Empirical Methods for Artificial Intelligence. The MIT Press, Cambridge, Massachusetts (1995)

Dalle, O.: On reproducibility and traceability of simulations. In: Proceedings of the 2012 Winter Simulation Conference (WSC) (2012)

Denney, E., Pai, G.: Tool support for assurance case development. Autom. Softw. Eng. **25**(3), 435–499 (2018)

Guardian Centers. Examples of previous full scale exercises. https://guardiancenters.com/exercises (2022)

Hawkins, R., Paterson, C., Picardi, C., Jia, Y., Calinescu, R., Habli, I.: Guidance on the assurance of machine learning in autonomous systems (AMLAS). https://arxiv.org/abs/2102.01564 (2021). Accessed 10 Nov 2022

Herrmann, J.W.: Data-driven metareasoning for collaborative autonomous systems. Technical Report, Institute for Systems Research, University of Maryland, College Park. http://hdl.handle.net/1903/25339 (2020). Accessed 10 Nov 2022

Houeland, T.G., Aamodt, A.: A learning system based on lazy metareasoning. Progr. Artif. Intell. **7**(2), 129–146 (2018)

Jarin-Lipschitz, L., Liu, X., Tao, Y., Kumar, V.: Experiments in adaptive replanning for fast autonomous flight in forests. In: 2022 IEEE International Conference on Robotics and Automation (ICRA). Philadelphia, Pennsylvania, 23–27 May 2022

Law, A.M.: Simulation Modeling and Analysis, 4th edn. McGraw-Hill, New York (2007)

Mesmer, D.B.: Validation framework for assuring adaptive and learning-enabled systems. Final Technical Report SERC-2021-TR-021, Systems Engineering Research Center, The University of Alabama in Huntsville (2022)

Molnar, S.L., Mueller, M., MacPherson, R., Rhoads, L., Herrmann, J.W.: Metareasoning to improve global and local path planning for a mobile ground robot. Technical Report, Institute for Systems Research, University of Maryland, College Park. http://hdl.handle.net/1903/29723 (2023)

Murphy, R.R.: Introduction to AI Robotics, 2nd edn. The MIT Press, Cambridge, Massachusetts (2019)

National Academies of Sciences: Engineering, and Medicine. Reproducibility and Replicability in Science. The National Academies Press, Washington (2019)

Nayak, S., Yeotikar, S., Carrillo, E., Rudnick-Cohen, E., Jaffar, M.K.M., Patel, R., Azarm, S., Herrmann, J.W., Xu, H., Otte, M.W.: Experimental comparison of decentralized task allocation algorithms under imperfect communication. IEEE Robot. Autom. Lett. **5**(2), 572–579 (2020)

NIST/SEMATECH: e-Handbook of Statistical Methods. http://www.itl.nist.gov/div898/handbook/ (2012). Accessed 10 Nov 2022

Open Robotics: Documentation. https://wiki.ros.org/ (2022). Accessed 6 Aug 2022

Polanyi, M.: Personal Knowledge: Towards a Post-Critical Philosophy. The University of Chicago Press, Chicago (1962)

Rabiee, S., Biswas, J.: IV-SLAM: introspective vision for simultaneous localization and mapping. In: Fourth Conference on Robot Learning (2020)

Raghupathi, W., Raghupathi, V., Ren, J.: Reproducibility in computing research: an empirical study. IEEE Access **10**, 29207–29223 (2022)

Reinelt, G.: TSPLIB. http://comopt.ifi.uni-heidelberg.de/software/TSPLIB95/ (n.d.)

Rous, B.: The ACM task force on data, software, and reproducibility in publication. https://www.acm.org/publications/task-force-on-data-software-and-reproducibility (2022)

Sung, Y., Kaelbling, L.P., Lozano-Pérez, T.: Learning when to quit: meta-reasoning for motion planning. In: 2021 IEEE/RSJ International Conference on Intelligent Robots and Systems (IROS), pp. 4692–4699 (2021)

Svegliato, J., Basich, C., Saisubramanian, S., Zilberstein, S.: Metareasoning for safe decision making in autonomous systems. In: 2022 IEEE International Conference on Robotics and Automation (ICRA), pp. 11073–11079 (2022)

Vincenti, W.G.: What Engineers Know and How They Know It. The Johns Hopkins University Press, Baltimore (1990)

Index

Printed in the United States
by Baker & Taylor Publisher Services